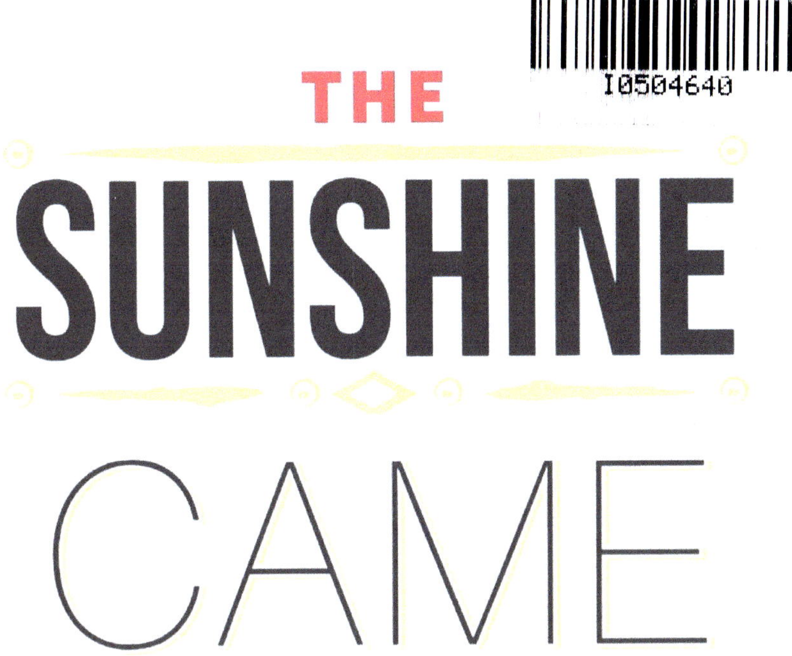

The Sun Never Sets But We Often Settle

Research | Essays | Sketches | Poetry

by

LLOYD JOSHUA SAMS

Foreword

Peace:

Dear Reader,

It is with great pleasure that I introduce to you "The Sunshine Came" by Lloyd Joshua Sams. This book is a testament to the power of hope and resilience, reminding us that even in the darkest of moments, the sun never truly sets. Through personal anecdotes and reflections, Sams invites you on a journey of self-discovery and encourages you to break free from settling for a mundane existence.

With its uplifting message and inspiring prose, "The Sunshine Came" will open your eyes to a brighter future, warm your heart with new beats, and expand your horizons to new heights. Sams' words will be a source of comfort and encouragement as you navigate life's challenges and obstacles. This book is a must-read for anyone seeking inspiration and a renewed sense of purpose.

So, sit back, relax, and allow Sams' words to guide you towards a brighter tomorrow. The sun never truly sets, and with this book in hand, you'll be reminded of the endless possibilities that await you.

Travel light,

The Sunshine Came

The Sun Never Sets, But We Often Settle.

By Lloyd Joshua Sams

The Birdwalker Publishing Company

"*The Sunshine Came* is an eye opening and groundbreaking perspective on how to find that piece of heaven within you no matter how gloomy days may appear. The sun never sets - but we often settle."

Dedication

"Jessica Rabbit, KVR"

(To the One that opened my eyes to see things for what they are and not necessarily what I want them to be.)

&

The Golden State, CA

(By far the most valuable commodity in the State of California is its scenery – the weather's not so bad either.)

Table of Contents

- Foreword
- Dedication
- Introduction
- Preface
- Chapter 1 – Balance
- El Coronado – The Crowned One (Sketch)
- Chapter 2 – The Sun is Always Present
- We are Standing (Poem)
- Chapter 3 – Keeping the Family Tree Alive
- Chapter 4 – The Importance of Sunlight…
- Making You Moonshine (Poem)
- Chapter 5 – Spark Up You Are A Powerhouse
- The Evolution of Revolution [Hooked on the Light] (Sketch)
- Chapter 6 – The Human Body | The Earth | The Sun
- Chapter 7 – The Beacon on the Mount
- Impartación "Impartation" (Sketch)
- Chapter 8 - Light of the World
- This Lil Lite of Mine (Poem)
- Speak it to Being (Sketch)
- Chapter 9 - The Apple Doesn't Fall Too Far From the Tree
- *Vida Inteligente "Intelligent Life"* (Sketch)
- The Sun, The Moon and The Stars (Poem)
- Chapter 10 – The Science of Light Bodies…
- Three Royal One Watcher (Sketch)
- Chapter 11 – We Are Seeds Planted in God's Garden
- Chapter 12 – Frequencies and Waves
- Chapter 13 – Travel Light

Table of Contents (Continued)

- Sunshine was His Name (Poem)
- Chapter 14 – Final Thoughts
- The Sun Reaching to Stars (Poem)
- Author's Note - The Sunshine is Coming Again

"Some are blind because they cannot see...

Others are blind because they do not see the light.

Still there are others who are blinded by the Light."

~ The Birdwalker

Preface

The Holy Bible is riddled with examples of relationships between men and women; but eventually all stories in the Holy Scriptures lead to one's relationship with their Creator.

On the path to finding the **Fruits of the Spirit** (Love, Joy, Peace, Forbearance, Kindness, Goodness, Faithfulness, Gentleness, and Self-Control) I have also found countering and polarizing **"Weeds of the Spirit"** such as: Addictions, Jealousy, Depression, Confusion, Control, and Suffering.

The "Weeds of the Spirit" create an overcast, blocking our Heavenly Shine, which causes our organic glow to emit an ambit of Light in a Dark world. We are born into darkness, so that we may be a lamp stand for others. You are a star.

Fruit is normally thought of as sweet, ripened, seed-filled, juicy, refreshing, natural, but mostly something edible and good for you, grown from the Earth; or from the ground – "the heartland" from which humans, plants, animals, and all living things derive their strength and sustenance – food.

Where does the dirt, soil and earth come from? *The Sunshine Came* will unlock esoteric and scientific evidence that we are all connected in an ecosystem of vibrations, consciousness, frequencies, and nature.

The author, in a basic way, informs the reader of their inherent ability to always shine with knowledge and wisdom - the greatest treasures one can attain in this life.

Deep in the abyss of melancholy, I found Peace and Hope in the crevices of disaster; there's much faith during chaos, especially, when all around you seems calamitous and destructive – yet you survive and thrive. There is also a term for this: *Post-Traumatic Growth.*

Introduction

Before I go about my day, I usually spend an hour or so clearing my mind by praying and meditating on positive and enlightening thoughts to focus only on pleasing my Creator. Often I include: Christian and Kundalini Yoga; Indian Vedic Mantras; and recite ancient Hebrew text from the Torah and Holy Bible to guide my actions on the Right path. I sometimes practice reciting these scriptures in other languages.

I'm not a big eater but I like to eat well! I normally do not eat a big breakfast, maybe a piece of bread and juice or milk.

Not trying to brag, but I can throw down in the kitchen. I mean you name it; I can cook it, grill it, roast it, bake, or BBQ it.

I'm from Texas. And in the South people are always eating and drinking something (mostly because our hospitality is the best in the world). Admittedly, New Orleans has some of the best food you'll ever taste, but you also may get your pocket picked along with the lint.

Genuinely, when people ask you how you feel in the South, they mean it. (Don't get it twisted no one wants to hear your life story either). I think it has something to do with the enriched spirituality of Southerners. Trust me; we pray an awful lot for rain AND pray for it to stop too. Buds will start to grow; cornfields and rose gardens will yield their harvest every year. Rinse and repeat.

When its dry and our crops need water, we do our "dance," and it rains. Then it starts flooding and we ask for it to stop. This book illustrates that we are like seeds encoded with exactly the genetic code that makes us unique individuals and planted exactly where we are supposed to be planted. The goal: is to blossom no matter where you're planted. Like growing pains, phases of growth are expected.

Because of planetary and celestial cycles, we are often prone to feeling the gravitational effects of polar shifts, alignments, eclipses, and oceanic currents that form who we are. And all of this takes place because of the force of the Sun.

Chapter 1
Balance

I've witnessed first-hand the beauty of life in harsh places or the beautiful side of ugly: kids in ghettos; the homeless in favelas in Brazil, and communities after tornadoes and hurricanes in the US Midwest. As a photographer I have taken pictures of beautiful cactus in the Southwestern United States' Desert Center which blossom into colorful vibrant flowers (reds, blues, whites) some grown out of dried volcanic petrified lava like in Mexico and South America with little to no hydration.

Life is always propagated upward toward the Sun. One thing I know for sure is that everyone has in common, is that the Sun has been present all the days of [our] lives. Actually, the saying "sunrise" or "sunset" is scientifically incorrect according to the proper alignment and rotational orbit of the planet and the stationary body of gases we call the Sun, which transmits light waves of energy onto us, feeding our personal, individual, and collective growths.

The Story of Jesus Christ shares with us the importance of Peace in knowing that through Him, eternal salvation is near, not salvation or eternal life forever with your "BFF," lover, or spouse, but through the Son. That is the link between all living things great and small - that through the son/sun we're a part of it All.

We have all been in a low place wanting to be risen upward and healed by something or someone. We often find ourselves caught up in worldly attractions leaving us still wanting unworldly "fruit" like the Fruit of the Spirit" described to us in the *Book of Galatians*.

More humans and living things on Earth are attracted to fruit than to weeds; to light rather than dark, but we sometimes find ourselves trespassing into wild thickets of thorny vines with no vision. Although herbs and spices are considered weeds, they're not. In most cooked meals herbs and spices are more than 10% of the overall total ingredients.

Can you imagine a recipe for a meal that requires 5 cups of sugar, 3 cups of black pepper and 2 cloves of garlic, oh yeah and a strip of chicken breast? Yuck! Who would enjoy eating such a meal? Further, I doubt it would be healthy to eat or good for the body. My point is that a meal must be balanced; an ounce of oranges is probably safer than an ounce of chili peppers at one sitting.

U.S. statesman and inventor Benjamin Franklin said, "an ounce of prevention is worth a pound of cure." After reading this book you will have a few pounds of prevention and an understanding of the world around you as well as a deeper understanding of your place in the world. Blossom where you are planted.

When you become enlightened you realize that you are rescuing your light forever as well as responsible for guiding others to their own shine. Although certain lights are not visible with the naked eye, light waves generate pulsation and vibration frequencies because light travels in every direction.

In order to be where the light source is, to charge our glow, we then have to travel as well in order to be constantly renewed and guided in the right direction through darkness.

When I got married to my first wife her family believed we wanted to jump the broom because she was pregnant; not because we loved each other. The truth is I deeply loved her. She was extremely smart, gifted at designing art, loved music from around the world and it didn't hurt that she was gorgeous. But things weren't balanced and in alignment between she and I.

Three weeks after we met, we decided to get married. Three months after that, we were hitched. Three years later, our son Joshua was born. To us, we did everything the way we were taught: Get school. Get degree. Get Family and then career, etc… I didn't have far-fetched ideas of white-picket fences or circle driveways and butlers, but the dream to attain more than our parents seemed to shrink year after year!

Increases in our income started to diminish at first until we started putting our ideas together and birthed the graphic design and music company. It was a one-stop-shop publishing boutique for indie artists and start-up companies.

Sooner or later, husband-wife businesses suffer from too much time working together instead of continuing to date and court each other. No matter how good money is or isn't, I learned that profits do not determine happiness in one's life.

Our joy could no longer be sustained by one another. I found joy in raising our son but not our union as in sharing in each other's individual passions. Instead, we faced an impasse, a stalemate situation; with neither of us willing to suffer long enough until the good started to outweigh the bad and drifted apart to divorce.

Whether the death of a loved one, or a divorce, an eclipse of the sun in one's life can be compared to a temporary darkening of the soul, where the light and warmth of hope and joy are obscured by the shadow of negative events or emotions. Just like a solar eclipse, these moments of darkness can be disorienting and unsettling, causing one to feel lost and uncertain. However, just like an eclipse, these moments of darkness do not last forever. The light and warmth of hope and joy will eventually return, shining brighter and stronger than before.

The significance of this comparison, light and dark, lies in the reminder that no matter how dark and hopeless a situation may seem, it is only temporary. The sun will always "rise" to be seen again, bringing with it the light and warmth of a new day; a message of encouragement, reminding us that even in the darkest moments, there is hope for a brighter future.

Therefore, we should not allow these eclipses to define us or dictate our future. We should hold on to hope, trust in God, and focus on the light that will eventually come. Just like an eclipse, these moments will pass, and the sun of hope and joy will shine again, bringing life and renewal to our souls.

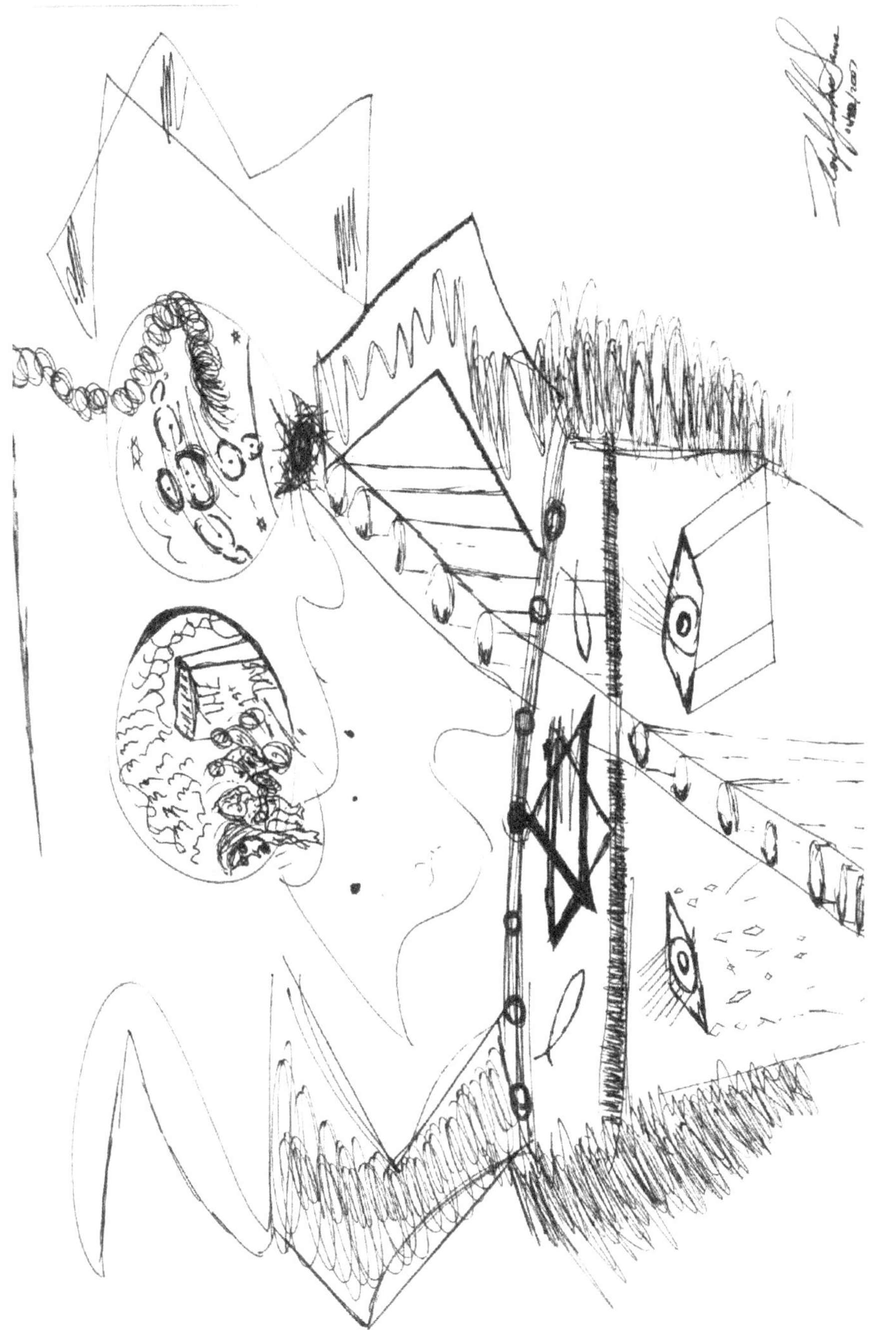

El Coronado (The Crowned One) **June 2007**

por El Ave Caminante (by The Birdwalker)

Chapter 2

The Sun is Always Present

The Sunshine Came is an outcry to people who have lost their voices in marriages; or made commitments to groupthink rather than individuality or those who do not know how to love themselves and others as God has loved us all. In the absence of Light, every type of wrong and misdeed can shun our blessings away. This scientific memoir provides a blueprint for connecting the dots to understanding one's existence which can lead to discovering purpose.

This book is a testimony and a teaching tool to assist you in seeing the chalice of life as half-full, instead of half-empty. The rays of the sun do not discriminate and are not empty, but packed with insight, wisdom, energy, food, information, and spirit.

People segregate themselves based on their own light-skin fetishes, dark fears, unidentified sicknesses, and poor attitudes. There is a power source greater than the sun and by the time you finish reading this book you'll learn how to identify this power source and how to plug directly into the surge of goodness without needing wires or a priest.

The Sun is the most important celestial body in our solar system. It is the center of the universe, around which all the planets revolve. The Sun provides warmth, light, and energy to the entire planet. Without it, life as we know it would not exist. In this chapter, we will explore what the world would be like without sunlight.

Firstly, without sunlight, the world would be completely dark. The Sun is the primary source of light for the Earth, and without it, we would be left with only the dim light of the stars and moon. This would have a profound effect on our daily lives, as many activities we take for granted would be impossible without adequate light. People would not be able to see well enough to read or write, and it would be difficult to travel, work, or carry out any other activities that require good visibility. Most of the globe would be in complete darkness.

The lack of sunlight would also have a significant impact on the climate. Sunshine is the primary source of heat for the Earth, and without it, the planet would rapidly cool down. This would lead to a dramatic drop in temperature, causing many regions of the world to become uninhabitable. Plants and animals would struggle to survive, and entire ecosystems would collapse as we have learned from fossils during the *Ice Age*. In fact, scientists estimate that the Earth's temperature would drop by an average of 50 degrees Celsius without the Sun.

Another consequence of the absence of sunlight would be the lack of photosynthesis. This is the process by which plants convert sunlight into energy, which they use to grow and produce oxygen. Without sunlight, plants would be unable to carry out photosynthesis, and the world's oxygen levels would gradually decline. This would have a catastrophic effect on all life on Earth, as oxygen is essential for survival.

In addition to the physical effects, the absence of sunlight would have a profound impact on human psychology. Sunlight is a natural mood enhancer, and lack of exposure to it can lead to feelings of depression and anxiety. Studies have shown that people who live in areas with less sunlight are more likely to suffer from Seasonal Affective Disorder (SAD), a type of depression that is triggered by the change in seasons.

It is worth considering the cultural and symbolic importance of the Sun. In many cultures, the Sun is revered as a deity or a symbol of divinity. It is often associated with warmth, vitality, and life-giving energy. Without the Sun, many of these cultural traditions and beliefs would lose their meaning and significance.

The absence of sunlight would have a profound and far-reaching impact on our world and galaxy. Without it, we would be left in darkness, struggling to survive in a cold and inhabitable environment. The loss of photosynthesis and oxygen would have a catastrophic effect on all life on Earth, and the psychological effects of the absence of sunlight would be equally devastating. The Sun is truly the life force of our planet, and we should be grateful for its warmth and light.

See, the sun is always present, but sometimes our lives are a little too cloudy to feel its rays. The nearer to God we draw, the closer we get to our destiny because we can see the laid-out path more clearly. Am I saying that the actual and physical sun itself affects the spiritual beliefs one has in God? Comprehensively, categorically, and cumulatively - Yes!

Cycles in weather, planetary orbits, and people pollution cause a multitude of effects on humankind, in our love lives, church lives, and our family lives. (I put Family and Friends in the same pool – "Framily"). I have heard that "religion was invented by man to control the masses." Whereby using gold leaf cathedrals and charismatic gestures to conquer lands and souls.

This book is not about religion; steps to self-help; how-to-hook-up, or how to keep your earthly relationships together as much as it is about learning to find the Light in your darkest situations.

Next to losing my Mother to cancer at 50 years young, getting a divorce was one of the worse things I could have ever encountered. Inevitably and unfortunately, it had to be done in order to keep me from what we thought was a dark and gloomy marriage riddled with lies and deceit.

Looking back with "20/20" (hindsight), we separated ourselves from the Light of the World and left the church because of past hurt and wrongdoings from members of our church who often seemed to carry a holier-than-thou air and undermined brotherly love.

Instead of asking for "reconciliation," we believed in "forgiveness" more, which was ill-fated when we did not truly mean it. Conviction leads to forgiveness. Forgiveness should lead to Repentance and then to Reconciliation.

The Final Step (Reconciliation) should offer a true and binding supernatural force that prevents offense or insult again toward the one you offended and forgave. There is no condemnation in the Light; all things done in the darkness shall be brought to the Light to resolve and reconcile. Need I remind you of the mistakes I made in my marriage in order to impress upon you that you, or I are not perfect - we all have regrets and skeletons in our closets and monsters under our beds - "booger-wolves" we haven't quite mastered. In bereavement there is comfort. Grief is the precious price for love.

For instance, in a tragic situation of a loved one, if it didn't happen to us we may be burdened with sadness and gloom but honestly, most people in the back of their minds are glad it didn't happen to them - and hope to make life-changes that bring more joy. Comforted by memories in the absence of loved ones - our brains are able to synthesize their voices; see their face(s); and even their scent with the connection of a Higher Power.

When I was a youngster, I remember girls playing *Patty-Cake* and singing *Sunshine You're My Sunshine:*

The lyric to Patty Cake detailed how to bake a cake in an oven as fast as she could by rolling the dough, patting, it and marking it with a "T." I later found out it was a nursery rhyme written in England in the late 1690's. The other song "Sunshine," I can only remember the first stanza:

You are my Sunshine
1 My only Sunshine
2 You make me happy
3 When skies are grey
4 You'll never know, Dear
5 How much I love you
6 Please don't take my Sunshine
away.

(1939, Jimmie Davis & Charles Mitchell)

Sunshine is the Official State Song of Louisiana made popular by Jonny Cash.

These songs stick out to me now because they are embedded with the fabric of harvesting earth's goodness and manufacturing a reason to be happy. The wheat and grain to make the flower to make the dough; to heat up the oven all takes the Sunshine and Light of the World to cast power to end famine, hunger, malnutrition.

I'm not only talking about physical nourishment, but I'm also talking about spiritual nutrition. A well-balanced emotional life is important for the mind as it is for the heart.

In the stanza I cited from the song *Sunshine* we see the writer compares the happiness one receives to the warmth and fulfillment of the sunshine even when "skies are grey." The final line suggests that it would be regrettable or not joyful if the Sunshine was taken away; concluding there is still a more powerful force that controls our experience to the fullness of the sun. Is it taken from us? Or do we move from its presence by burying our heads in the sand?

We are Standing

A family of trees, a family affair,

A family of ties and Vanity Fair.

Roots we share; fruits we bear –

Hands we hold, at the beginning of prayer.

Amen. We stand!

Amen. We stand!

Some dance, some sing, some play

Some walk, some ring, some stay.

Some go, some bring, some take,

Some give, some live, some die - yet

We are standing!

We are standing!

~ Lloyd Joshua Sams (08/25/2002)

Chapter 3

Keeping the Family Tree Alive

When I was a boy, I remember spending the summers with my grandmother and my dad's side of the family on the island of Galveston off the coast of Texas. I can still hear my Uncle Roy whispering in my ear "It's time to get up, the fish and crabs are biting' - we gotta hurry before high tide."

I don't remember how early in the morning it was, but the sun hadn't come up yet and I could smell my grandmother's Community coffee brewing in the downstairs kitchen.

Uncle Roy would go on and on about it being the second full moon in July and how the crabs were gonna be stuffed full because of the *blue moon*. I didn't get it then but what he was talking about was the cycle and seasons of fishing, hunting, and survival:

One - You gotta get up early.
Two - You gotta know when to move.
Three: If you don't work you don't eat.

These fundamentals were very important. I'm not saying that I use these three principles to guide my every move but, in all things get understanding. What I now understand is that patience, persistence, and a grumbling stomach will make you get up early as does the "Early Bird."

My family wasn't starving.

For months I worked in an office with no windows. In the Fall / Winter months I would get to work before the sun rose and leave near sun set. I couldn't wait to take a break or go to lunch outside. At first it would take a few minutes for my eyes to adjust because of the brightness of the sunlight, but I could immediately feel the reviving warmth despite a chilling cold temperature.

Often, people have no idea what goes into farming or agricultural development. The proper and appropriate actions to yielding a harvest requires "manpower," man-hours, and know-how. The know-how comes through experiences and apprenticeship. Most farms are tended by third or fourth generation bloodline support within the family (like the Walton Family).

Sure, facts can be found in the Farmer's Almanac and can offer many techniques and methods for planting and growth certain crops. However, sweat-equity, resources, time, care and patience are also required.

Not every seed will take root in certain fields, terrain, geographic region, or climate. Some seeds require much water and others the slightest bit of moisture will produce eye-stunning flowers with little to no attention. However, all plant life on earth requires sunlight in ORDER to grow.

Our own family trees will not grow correctly without the proper Light in our lives. Newborn babies require sunlight to receive nutrients like Vitamin-D to avoid illnesses such as jaundice. Confinement without light is considered cruel and unusual punishment to the point that our nation's worst criminals in America are obliged to at least 30 min of sun a day.

Other than the supreme force in this world, the Sun is also the most required provision to live and maintain balanced lives. Ordering and organizing our life should prioritize our days, hours, minutes, and seconds of our time here. Extracting the best from every situation no matter how gloomy it appears.

The same pilot my grandmother used to light her kettle is the same Pilot that guides the fish, crabs and, even us to the shoreline. The Light of the Sun which casts its reflection onto the Moon gives us the glimmer to feed ourselves. Tantalizing sea life - all life - on land or in the air is attracted to the Light of the World or the Sun.

Homosapien sapiens have traveled miles, light years, and centuries to understand all that exists within the power of the Sun. We are driven by the sundial as fish are drawn to bait or crabs to a net with chum.

We all get caught up, hooked on the line, or even worse get thrown back in the waters to find our way back home like troupes of salmon flipping their way upstream against the current like Jonah being spewed out onto land. Our time is spent like a mousy Ferris wheel trying to keep up with the *Jones'*.

We're all going somewhere, no doubt. We're all going at some time. Some will come here. Some will go there. We were all born into darkness to find the light. We are designed to follow the Light even when we attempt to hide ourselves from the Sun.

We cannot dismiss the ebbs and flows of life. There are many bodies or ships you will find yourself riding in as a passenger along this eternal journey to Heaven borne through the heavens on earth and beyond.

There is nothing new under the sun.

When high or low tides come; when typhoons and tsunamis hit our lives; when we find ourselves to be crabs in a bucket or tangled in a net or network of the wrong type of friends or people that don't have your best interest at heart, nor the capacity to facilitate growth in a higher direction - **we must yield to the power of the Son to calm our waters, wash away our doubt, fear and dry our tears**.

On a solid foundation a temple can be built to withstand the calamities of life and Acts of God. Mansions and vessels are already commissioned to carry us through the storm and to our destiny with the Reigning Champ of All Time - the Ancient of Days.

"It takes 2..." Who hasn't gotten married (or at least tried)? I can't be the only adult person in the world who thought I found God in a person I loved dearly.

It takes two; just not the two you probably have in mind.

Eventually, *The Sunshine Came.*

Chapter 4

The Importance of Sunlight - Biopsychosocial View

Sunshine is essential to human life and well-being.

Our bodies are naturally wired to the sun and the cycles of the universe, which dictate our *circadian rhythm* and regulate our sleep patterns. The sun provides us with warmth, light, and energy, and has been credited with numerous health benefits, including improved mood, enhanced immunity, and increased mental alertness.

According to the Merriam-Webster Dictionary, the definition of the term *biopsychosocial* is or of, relating to, or concerned with the **biological**, **psychological**, and **social** aspects in contrast to the strictly biomedical aspects of disease or dysfunction.

From a biological perspective, sunlight is the primary source of Vitamin D, which plays a critical role in maintaining strong bones and a healthy immune system. Sun exposure also increases the production of serotonin, a hormone that regulates mood and sleep.

The Effects of Sun Deficiency:

Unfortunately, many of us now spend most of our days indoors, away from the sun's nourishing rays. As a result, many people suffer from a deficiency in Vitamin D, which can lead to a range of health problems, including osteoporosis, weak immune systems, and a greater risk of developing certain types of cancer.

In addition to physical health problems, lack of sun exposure can also have a significant impact on our mental and emotional wellbeing. Studies have shown that sun deficiency can lead to a myriad of feelings of fatigue, depression, and anxiety, as well as disrupted sleep patterns and decreased mental clarity.

The Benefits of Outdoor Living:

To maximize the benefits of sunlight and prevent sun deficiency, it is important to spend time outside in the sun on a regular basis. This can include activities like gardening, hiking, or simply lounging in a park.

Being outdoors and absorbing sunlight helps us to grow and thrive, much like plants. Not only does sunlight provide us with Vitamin D and other essential nutrients, but it also stimulates the production of endorphins, which are natural mood boosters that can help to improve our overall wellbeing.

The sun is a critical part of our lives and our health, and we need it to thrive. Whether through intentional sun exposure or simply spending more time outdoors, incorporating the sun into our daily routines can bring numerous benefits to our physical, mental, and emotional health.

The Social Benefits of Sunlight:

In addition to its physical and mental health benefits, sunlight also plays an important role in bringing people together. As social beings, humans thrive in environments that provide opportunities for connection and community, and the sun provides just that. Outdoor events and activities, such as picnics, festivals, and beach trips, provide opportunities for people to come together, share experiences, and strengthen bonds.

In areas where sunlight is scarce, such as Seattle or Ireland, rates of depression and suicide tend to be higher compared to areas with more abundant sun exposure. This is because a lack of sun exposure can lead to feelings of isolation and hopelessness, which can contribute to the development of mental health problems like depression.

The Impact of Sunlight on Mental Health:

The impact of sunlight on mental health is well documented and widely recognized. Studies have shown that exposure to sunlight can help to improve mood, reduce anxiety and depression, and increase feelings of happiness and well-being. This is why many people find that spending time in sunny climates can help to relieve feelings of stress and anxiety, and why light therapy is often used as a treatment for Seasonal Affective Disorder (SAD).

One possible explanation for these effects is that sunlight regulates the production of hormones like serotonin and melatonin, which play a critical role in regulating mood and sleep patterns. Sun exposure also stimulates the production of endorphins, which are natural mood boosters that can help to improve our overall wellbeing.

The sun is an essential part of our lives, both for our physical and mental health, and for our social well-being. By incorporating sun exposure into our daily routines, we can reap the numerous benefits of this natural resource, from improved mood and enhanced immunity to stronger relationships and greater overall happiness. So, next time you have an opportunity to spend time in the sun, take it - your body, mind, and soul will thank you!

Makin' Your Moon Shine

As the world turns, I spin your Pluto.
Makin' your Moon shine, intoxicating myself,
With Jupiter's red velvet wine.
We zoom pass Neptune, as I make your Mercury rise.
0-G. No gravity as we surpass The Ancient's skies.
I pass through Saturn's rings,
Like a voice from a soul that sings.
Makin' your moon beam -
We curve straight lines, around Mars.
Whistling fast pass stars, as if no one had seen us –
Except though, the faithful morning stare of Venus.
The moonshine flows between your toes,
Inebriating your spirit with each of my blows,
Down your neck; your back; and your Uranus too!
And nine months of fermentation,
Birthed 9 planetary shots of Moon Shine,
For you to sip like ethereal fine wine,
And come, come, come, like you have before...
Back to Earth through the Sun's door.

~ Lloyd Joshua Sams (09 / 15 / 2002)

Chapter 5

Spark Up - You are a Powerhouse

There is a spark in us all. Some call it the divine particle; the Holy Spirit; a piece of God the Creator. Universally, we are entangled in the etheric web of pathways to our Original selves. Born into a cruel and dark world, we innately yearn to receive the power and love that generated us to be, in the first place.

Love is power. God is Love. We become lamp stands of Joy when we fuel ourselves with the Light of Love-Power. Within us all, is a light to shine:

This little light of mine
I'm gonna let it shine
X 2
And…Everywhere I go
I'm gonna let it shine
Let it shine – let it shine

Nearly every Christian child is taught this hymn growing up. It is probably one of the first songs ever learned next to the Happy Birthday Song and A-B-C's. As a child I remember singing it boldly and proudly. Even though it stuck with me all this time, it has been difficult to "let [it] shine" as it did when I was a boy. Do you know this song? Do you remember singing it as a child? When was the last time you sang it? How did it make you feel? What happened as a result of singing it?

When I grew up the phrase: "getting your shine on," "rise and shine," and phrases like "you're bright (smart)," and "You are beaming with confidence," were remarks of the highest compliment by someone to another person.

Girls liked glitter and sparkly things; Boys loved trying to create light by playing with fire – they both love *Northern Lights* in their minds and the first sighting of a troupe of fire flies during summer in the fields amid castles in the clouds; bon fires ricochet ambers through the wind as lovers' eyes catch their flight of fancy unto the sparkle which lands between two souls of two people which if the attraction is sexual in nature, will lead to a chemical reaction fluttering the heart and begin to percolate thoughts of ecstasy.

"The meeting of two personalities is like the combining of 2 or more substances - if there is a reaction, both are changed forever."

~ Karl Jung ~

It is my opinion that ecstasy is hard to be imagined with no light, although, in darkness it can be easily stimulated. What do I mean by this?

Artificial light from invention does not provide the same source of goodness to our lives as does natural light from the sun. Of course, too much of anything is not balanced and not living a balanced life isn't healthy. Too much imagining and ecstasy and not enough natural light will lead to a depressing and artificial life not connected to the source of true happiness - the Light of the World; an "oasis of love in a troubled world" (Osteen, Sr., 1989). Job said, "Everything on earth is a war," and the greatest of all battles is in our mind.

...Back to "getting your shine on." Women and men alike try to cover their unique marks or "blemishes" with make-up, enhancers, or even physically altering surgeries and injection procedures. I'm not advocating no make-up or that men and women stop grooming. I am saying that we should celebrate our unique differences.

Our grey hair from - when days in the sky above was gray; Our moles and birth markings put on us like the Baker's "T" (origin – The Cross, a symbol of Eastern migrant Orthodox Coptic Christians that was tattooed or marked on their hand or arm, a cross symbolizing their lineage to Jesus Christ) - the "t" (has to be lower before we are able to become upper: from t to T – the Light of the World is the Baker.) The difference between HERE and THERE is the cross.

Alpha, The Omega, has been there and done that.
(Please read John 6:26 – 58)

Look, no matter where a person is on the map, we all ask God at some point to reign and rain blessings in our lives because we know we don't have the power to rinse away all our sins or enough tears to flood away our worries and concerns. We draw near and do our dance; as promised, we see the Light, and begin to think that the glimmer of hope is the final spark…
It's not the final spark.

There have been several scientific studies conducted to explore the idea that a human skin cell can store information. These studies have focused on the storage capacity of DNA, which contains a unique genetic code that is embedded in the formation of every human, animal, and seed.

One study, conducted by researchers at the University of Washington, found that a single gram of DNA could store up to 215 petabytes of data. This is equivalent to the storage capacity of 170 billion iPhones. The researchers achieved this by synthesizing a DNA sequence that encoded digital data, which was then sequenced and decoded to retrieve the original information.

Another study, published in the journal of *Science*, found that DNA could be used to store and retrieve multiple copies of the same image. In this study, researchers used a process called *DNA-PAINT* to store and retrieve images of a cat and the Mona Lisa. The researchers found that DNA was a stable and reliable storage medium, and that it could be used to store and retrieve large amounts of information.

A third study, conducted by researchers at Harvard University, demonstrated that DNA could be used to store a book. The researchers encoded a 53,000-word book into DNA and then sequenced and decoded the DNA to retrieve the original text. The researchers found that DNA was an efficient and reliable storage medium, and that it could be used to store large amounts of information in a small space.

These studies demonstrate that DNA is a powerful storage medium that can be used to store and retrieve large amounts of information. While the storage capacity of a single skin cell is much smaller than the amount of DNA used in these studies, it is still significant. According to one estimate, a single cell of human tissue can store up to 1.5 megabytes of information. Your body is encoded with the powers of your own imagination, your ancestors, and your Creator. Every birth is a miracle.

The methods used in these studies to demonstrate the storage capacity of DNA involve synthesizing a DNA sequence that encodes digital data, and then sequencing and decoding the DNA to retrieve the original information. These methods are based on the fact that DNA is a specific and unique code embedded in the formation of every human and animal. The unique sequence of nucleotides in DNA provides the code that is used to create proteins and other cellular components, and this code is passed on from generation to generation.

Scientific research has shown that DNA is a powerful storage medium that can be used to store and retrieve large amounts of information. While the storage capacity of a single skin cell is much smaller than the amounts used in these studies, it is still significant. The unique genetic code embedded in DNA provides the basis for the storage of information, and research into this area is likely to continue in the future.

Where did this embedded coding come from? Who composed the sequence we call DNA? What can we do with this knowledge?

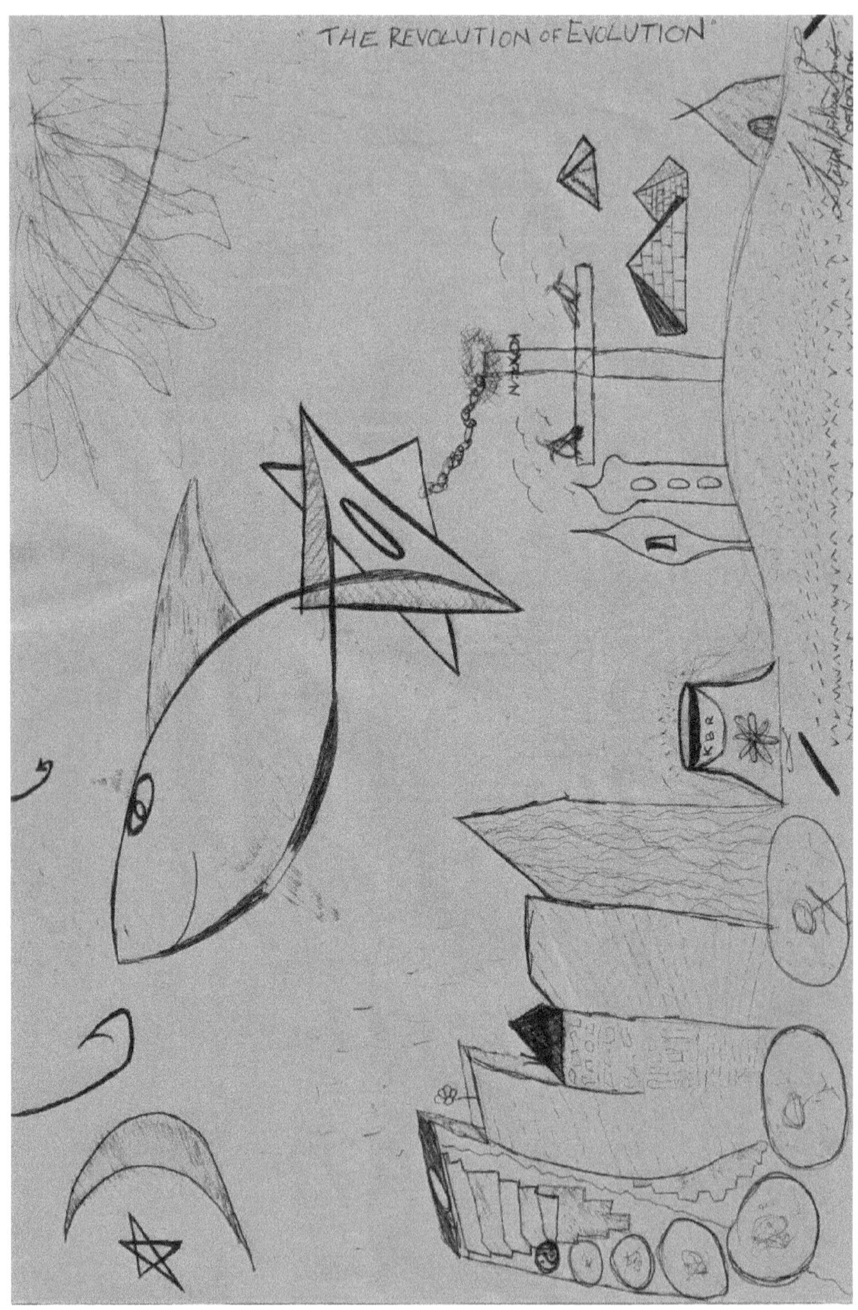

The Revolution of Evolution

[Hooked on the Light] August 2006

By The Birdwalker

Chapter 6

The Human Body | The Earth | The Sun

The Connection between the Solar Plexus and the Sun:

The connection between the sun and human health goes beyond just physical and mental health. In the field of holistic health and spirituality, the connection between the sun and the human body is believed to extend to the chakras in the body, the energy centres that run along the spine.

Of these chakras, the solar plexus chakra is especially connected to the sun and its energy. The solar plexus, located in the abdominal area, is associated with the power of self, personal will, and the ability to manifest one's desires. When this chakra is balanced and energized, individuals tend to feel confident, empowered, and in control of their lives.

According to holistic healing practices, the sun's energy can be harnessed to balance and energize the solar plexus chakra. This can be achieved through practices like meditation, visualization, and spending time in the sun. By doing so, individuals can experience greater clarity, self-confidence, and a sense of personal power.

The Connection between the Earth and the Chakras:

In addition to its connection with the sun, the solar plexus chakra is also connected to the earth and its energy. The earth is a grounding force that provides stability and balance which is seen as a source of nourishment and renewal.

Just as the sun provides energy and life to the solar plexus chakra, so too does the earth provide energy and stability to the root chakra, which is located at the base of the spine. The root chakra is associated with our sense of safety, security, and connection to the world around us. When this chakra is balanced and energized, individuals tend to feel grounded, stable, and in touch with their surroundings.

By connecting with both the sun and the earth, individuals can create a harmonious balance between the physical and spiritual realms, allowing them to tap into a greater sense of well-being and inner peace.

The sun, the earth, and the human body are all interconnected in complex and fascinating ways. Through practices like meditation, visualization, and spending time in the sun and nature, individuals can tap into this connection and harness the energy and life-giving properties of these celestial bodies to improve their health and well-being. Whether you are seeking physical health, mental clarity, or spiritual connection, the sun and the earth offer an endless source of nourishment and renewal.

The sun's rays are far reaching and ever-present like the omnipotence, omniscience, and omnipresence of the Light of the World.

The human body, the sun, and the earth are interconnected and related in numerous ways. The sun is the source of light and heat that makes life possible on earth. Without the sun, life as we know it would not exist. The human body, like all other living organisms, depends on the energy and nutrients provided by the sun for survival. In this essay, we will explore the various ways in which the human body, the sun, and the earth are connected and related.

The sun provides the earth with heat and light energy that is essential for the growth of plants, which is a critical source of food for human beings.

Photosynthesis is the process by which plants use sunlight to convert carbon dioxide and water into glucose, which is a form of sugar that the plant can use for energy. In turn, humans consume these plants for nourishment, and the energy they provide is used to fuel our bodies.

The Sun and Its Connections with the Body

The sun also plays a vital role in maintaining the body's internal clock, or circadian rhythm. This rhythm helps regulate the sleep-wake cycle, hormone production, and other physiological functions. Exposure to natural sunlight during the day helps keep the circadian rhythm in sync, which is essential for maintaining good health.

Furthermore, the sun is a vital source of vitamin D, a nutrient that is essential for healthy bones and overall well-being. The human body produces vitamin D when exposed to sunlight. However, many people do not get enough sunlight exposure to produce adequate amounts of vitamin D. This is especially true for people who live in areas with long winters or those who spend most of their time indoors. As a result, many people may need to take supplements to maintain healthy levels of vitamin D.

The earth's atmosphere plays a crucial role in protecting the human body from the harmful effects of the sun's ultraviolet (UV) radiation. The ozone layer in the earth's atmosphere absorbs much of the sun's UV radiation, preventing it from reaching the earth's surface. However, the ozone layer is being depleted by human-made chemicals such as chlorofluorocarbons (CFCs), which can have harmful effects on human health. Excessive exposure to UV radiation can cause skin damage, eye damage, and even skin cancer.

The earth's magnetic field also plays a role in protecting the human body from harmful solar radiation. The magnetic field acts as a shield, deflecting most of the charged particles that make up solar radiation. However, the magnetic field is not perfect, and some solar radiation can penetrate it, leading to harmful effects on human health. Solar flares, which are eruptions of charged particles from the sun's surface, can disrupt the earth's magnetic field, leading to increased radiation exposure.

The human body, the sun, and the earth are interconnected and related in numerous ways. The sun is the source of life on earth, providing the energy and nutrients that are essential for human survival. The human body, in turn, depends on the sun for maintaining the circadian rhythm, producing vitamin D, and maintaining overall health. The earth's atmosphere and magnetic field play crucial roles in protecting the human body from the harmful effects of solar radiation. It is essential to understand the relationship between the human body, the sun, and the earth to appreciate the importance of maintaining a healthy balance between them.

The Stress Gene - Cortisol

The purpose of cortisol is to regulate energy and metabolism. This is the sense or system in the body that is not mentioned in the 5 (or 6th or 7 senses we commonly discuss in human health). The "Flight-or-Fight Response System" is also known as the "Sink or Swim" affect which is actually a physiological sensory perceptive trait. This is a survival mechanism that has been developed in the body to ensure the greatest possible chance to "living to see another day."

***There is an unusual quagmire involving this gene, however.**

There is a drawback to cortisol. It breaks down muscle tissue and suppresses the immune system. Here's an example: If a tiger were chasing you, you wouldn't need protein, fat, or an immune system, you'd need lots of energy to escape. A quick source of energy is provided by the cortisol gene yet is detrimental to other systems in the body. Too low or too high levels of Cortisol can negatively affect the body.

If too high: fatigue, insomnia, decreased sex drive, erectile dysfunction, sugar and salt cravings; depressed mood; anxiety; irritability; bone and muscle loss. Either, hi or lo, can lead to serious heart disease if overused. Your body produces cortisol when you are stressed or being thrust into a life-or-death situation(s).

Like a seashell at a beach; tree rings; and even the human ear is formed in a pattern that reflects the rotation of the earth around the sun. The human ear is actually a symbolic pattern of what a person looked like as a fetus in the first trimester. Just as a star is seen from planet earth, the light that is emitted from the combustion of gaseous matter sparks a beam of light that one will observe, now, "light years" from the past.

Phantom memory is a condition in which a person who loses a limb or bodily faculty and cannot distinguish from missing that part when one thinks of using it to accomplish a task. Basically, even though the individual knows they're missing a body part, the muscle memory that causes neurotransmitter synapses to certain parts of the brain to move that "phantom" or ghostlike body part is activated even though it does not exist.

Today, with the prevalence of wounded soldiers returning from war, new scientific research in neurological experimentation is proving that the brain will develop and create new neuropaths to facilitate new movements to assist the body in compensating for the loss of a limb.
This often frustrates a paraplegic or quadriplegic in the beginning until they are able to overcome the disorder and begin to regain control of their muscular abilities.

At times I know it seems that heaven is extremely far away and that God of the Universe; the Great Architect is seated distant from us. But no matter where you are in the galaxy, you're never far from God - The Creator, because he is the father of Creation and of All Life.

Chapter 7

The Beacon on the Mount

"A fisherman in 2014 caught a see-thru fish off the coast of New Zealand. Scientists have not yet classified the unclassified species yet have linked it to the current day jellyfish, however, no such specimen has ever been recorded. The translucent body of this fish, marine biologists believe that it is an evolutionary genetic trait to allude predators and take advantage of light to reflect in its environment as in a camouflaged adaptation."

It is said that in the holy scriptures of the Bible in the New Testament that Jesus Christ at Mount Sinai demonstrated before his disciples a very special moment. Jesus shared a parable with his team as he so often did, but on this rare occasion he shown himself to his staff.

He unrobed his cloak and bore his body bare. He opened his chest and casted onto the men a window to heaven, transfiguring himself into a beam of radiant light. Jesus' moment of transfiguration was a sign to his followers that He indeed was filled with Light and not from this world. As hard as this may be to believe, it may be harder for you not to believe:

"Ye are from Beneath; I am from ABOVE. Ye are of this world; I AM NOT OF THIS WORLD" - John 8:23

"No one, when they have lit a lamp, puts it in a secret place or under a basket, but on a lampstand, that those who come in may see the Light. The lamp of the body is the eye. Therefore, when your eye is good, your whole body also is full of light. But when your eye is bad, your body also is full of darkness. If then your whole body is full of light, having no part dark, the whole body will be full of light, as when the bright shining of a lamp gives you light." -

Luke 11:33 - 38

By now I hope to have established for you the correlation and relationship of the Son / Sun, light, and the spiritual effects in our physical "real-world" lives. In case you are still unsure about this significant finding continue reading:

The words of Jesus spoken in Luke, Chapter 11 are very powerful and force any Believer to accept the "Light" doctrine. It is not my doctrine it's the Law under which all natural laws succumb.

The Mt. Sinai Moment of Transfiguration and this last scripture is paramount in understanding that we too are all marked by our God like a car manufacturer places their symbol on a new brand or model - your Creator placed his seal on you and has called every one of us to shine brightly. (Please read: Malachi 3:3)

God is the Baker; we are the dough. He did not make us as fast as he could. In taking time to contemplate the heavens and the earth he thought about the beginning and the end.

The best baker in town always uses the best ingredients, spices, flowers, and takes time sifting the flour, mixing the batter and heating the stone-oven to the Perfect degree - before the bread is ready - just as the yeast starts to rise, the whole town can smell the dough from blocks and miles away.

The wind carries news of fresh goodness across the Baker's land. Those who help harvest the grains eat for free; and the bread of life continues to feed those under the sun. Here are additional scriptures to read: Mathew 17:2, 5:13 - 16, Isaiah 52:1, Malachi 4:2, Ezekiel 28:15, Romans 13:14.

The sun is not just a celestial object in the sky, but a symbol of God's connection to humanity. The sun is a powerful and transformative force, shining with the light, warmth, and life that God provides to His creation.

The sun plays a significant role in many passages throughout the Holy Bible, symbolizing the power, grace, and presence of God. Here are 12 moments in the Holy Bible where the sun is mentioned, and its miraculous presence benefited mankind.

Each of these 12 moments in the Holy Bible where the sun is mentioned represents a unique and significant aspect of God's relationship with humanity:

1. Genesis 1:16 - God created the sun, moon, and stars to rule over the day and night, **showing His sovereignty over all of creation.**
2. Joshua 10:12-13 - The sun standing still in the sky shows God's intervention in the affairs of humanity, **demonstrating His love and protection for His people.**

3. 2 Kings 20:8-11 - The shadow on the sun dial going back ten steps is a sign of God's power to alter the natural order, **showing His ability to work miracles in the world.**

4. Isaiah 60:19-20 - The sun being replaced by God as the source of light and glory represents the ultimate fulfillment of His promise **to bring light and life to humanity.**

5. Ezekiel 32:7-8 - The sun going down and the land becoming dark is a symbol of God's judgment and destruction, **reminding us of His power and justice.**

6. Malachi 4:2 - The sun of righteousness rising with healing in its wings is a promise of God's salvation and restoration, **bringing hope to the world.**

7. Matthew 17:2 - Jesus being transfigured before his disciples with his clothes becoming dazzling white as the sun, **represents His holiness and the manifestation of God's glory in the world.**

8. Mark 1:32 – The sun setting when Jesus healed a man with an unclean spirit represents the arrival of God's healing power in the world, **bringing salvation to those in need.**

9. Mark 13:24-25 – The sun being darkened and the moon not giving its light is a sign of the end of the world, **reminding us of God's ultimate power and control over all things**.

10. Luke 1:78-79 – The rising of the sun on those who lived in darkness **represents the coming of Jesus, the morning star, bringing light and life to those who are lost in darkness.**

11. Acts 2:20 – The sun being turned to darkness and the moon to blood **represents the final judgment of God, reminding us of His justice and power.**

12. Revelation 7:16 – The sun no longer being scorching and no more need for shade represents the end of pain and suffering, **showing God's ultimate plan to bring His children into His eternal and loving presence.**

Through the sun, God shows His sovereignty, power, love, and salvation, reminding us of His presence and work in the world. These passages show that the sun is not only a physical object in the sky, but also a symbol of God's power, grace, and presence. Whether it is standing still in the sky or shining with healing power, the sun is a reminder of God's miraculous and transformative work in the world.

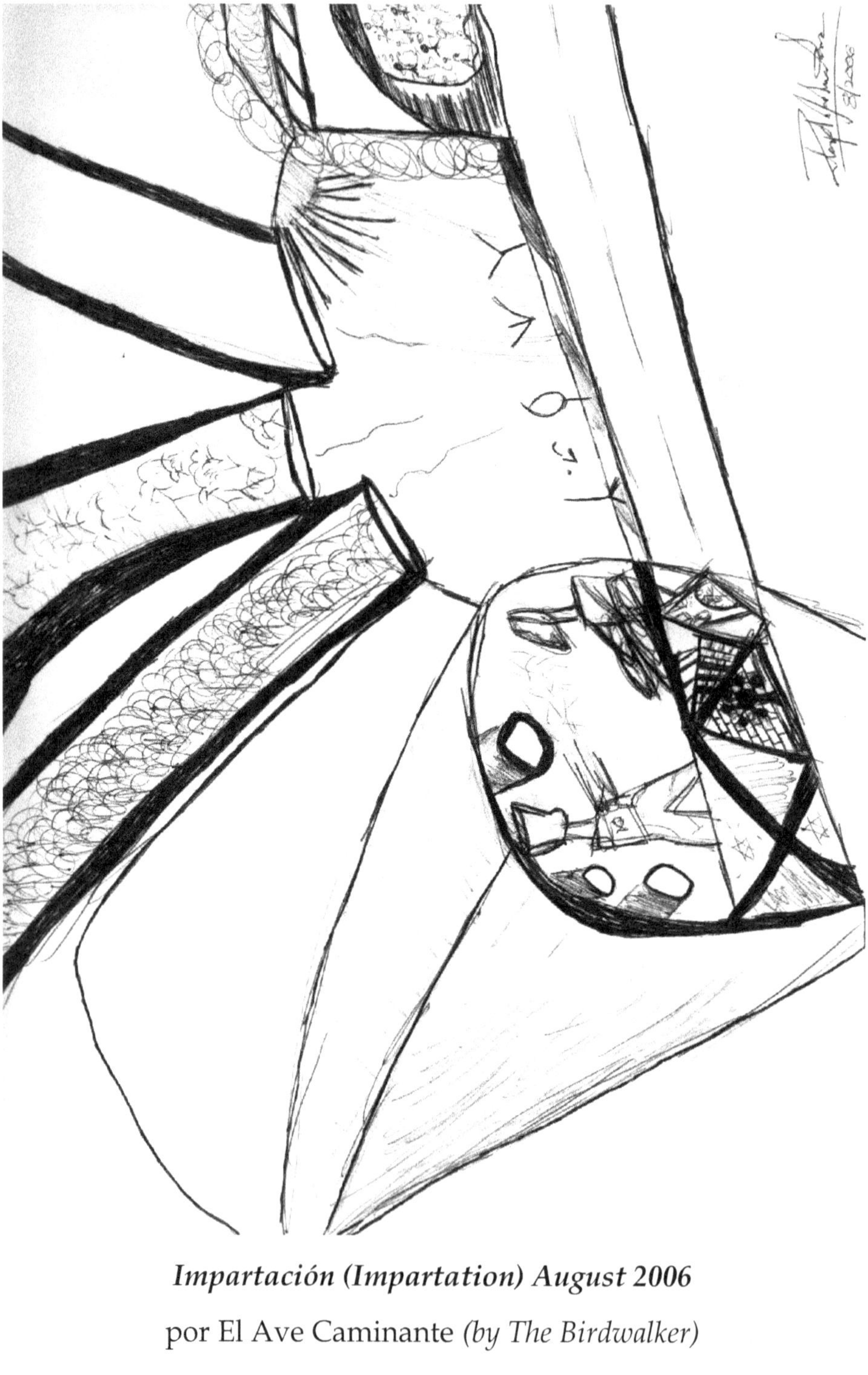

Impartación (Impartation) August 2006

por El Ave Caminante *(by The Birdwalker)*

Chapter 8
Light of the World

It is our duty to love ourselves in relationships as well as the subjective noun we are consciously observing. We can measure how well we love ourselves by looking at how well we treat those around us, especially family and friends.

All people; places; things; and ideas should be respected in this light. When we are sensitive to the existence of a particular stimulus, then we are giving credence and observation. When you observe something, you share its power and act as a vessel or capsule to contain a piece of its purpose. Everyone is but a small piece of God's plan, however, we each carry equal importance.

The weights we anchor in our lives are by choice; the more we observe the Son of Man we are able to gain strength to free ourselves from the shackles of regret, depression, loneliness, defeat, poverty, abandonment, and mediocrity. The greatest relationship one can have is with the Light of the World the "Christ" - Jesus!

The *Three Magi* (or priests), wise men were sent by King Herod and charged with following a resilient star in the Northern sky. Fearful that the orb of light was a sign that his kingdom would reign no more based on prophecy, King Herod decreed that all first-born males under two be killed as to stop God's plan of incarnating from the Master of the Heavens to the physical body of the long anticipated One: The Messiah.

Of course, we know that the 3 Wise Men followed the star floating through the sky over Israel. Eventually the radiant shining orb hovered over a shed in the desert where Mary, Joseph, and the Light of the World - Jesus, who lay mercifully under the watch of Father in heaven.

As foretold by many prophets the arrival of our deity figure in the flesh, the Son of God is now then gifted lavish jewels, presents, and fragrances as a gift from Herod, however, also instructed the Magi to return with detailed information of Christ's whereabouts, so he too like the other first-born males would be murdered as to prevent the fall of his menacing rule over the tribes of the Jews.

After honoring, anointing, and blessing the young Savior who will eventually die to rectify and reverse our transgressions with our Creator and magnify our power here on earth by bestowing the greatest Gift of All - His Son in the Flesh, Jesus! Through Jesus, the power He accessed in heaven from His Father maker of the Greater Light, the Sun shared with all Believers the light of the Holy Spirit which has been present always - since the beginning when - "God's spirit hovered over the face of the waters." Our Creator's very first decree was to make light, create light, and form light to be placed inside each of us, and in all things.

<u>Scriptures in the Holy Bible that prove the Sunshine Came theory based on the Light of the World Principle</u>:

1. **1 John 1:5**
2. **1 John 17-9**
3. **1 Peter 2:9**
4. **1 Thessalonians 5:5**
5. **2 Corinthians 4:4-6**
6. **2 Corinthians 4:6**
7. **Acts 20:18**
8. **Acts 26:13**
9. **Acts 9:3**
10. **Daniel 2:22**
11. **Ecclesiastes 2:13**

12. Ephesians 5:14
13. Ephesians 5:8
14. Exodus 13:21
15. Genesis 1:13-19
16. Genesis 1:3-5
17. Isaiah 42:16
18. Isaiah 45:7
19. Isaiah 60:1
20. Isaiah 60:20
21. James 1:17
22. Job 38:19-24
23. John 1:4-5
24. John 12:35-6
25. John 14:6
26. John 3:19-28
27. John 8:12
28. Luke 11:34
29. Matthew 5:14-16
30. Philippians 2:15
31. Proverbs 20:27
32. Proverbs 6:23
33. Psalms 104:2
34. Psalms 119:105
35. Psalms 119:130
36. Psalms 8:3-4
37. Revelations 21:23

This Lil' Lite of Mine

My spirit fits between the winds,
Like my soul within my skins.
My shadow; my sidekick - my friend,
My DNA climbs the ladder to the dimple in my chin.
My wisdom keeps me from being a "fool rushing in"
And the knowledge to speak Spanish, Chinese, and Russian
I have faith in God,
So with man there's never a question.
My heart is my body's mechanical engine.
I'm mixed with African, Creole, and Injun;
In other words, Negro, Frenchman, and Indian…
This Lil Lite of Mine,
I'm gonna let it shine to the end.
This Lil Lite of Mind,
I'm gonna let it shine from within.
Adam was here first,
But Eve was first to sin.
Thanks to Abba Abraham,
We're all one as blood Kin.
My fingertips are skilled like blacksmiths and artisans;
Like the Carpenter who healed with his hands.
Doctors don't cure, they feed you medicine.
You keep taking them even though you're hesitant.
O, but it eases the pain,

Hydro-Codone or Vicodin,

What's man made will leave you buried in the shade,

When the flesh is gone your bones begin to decay

Your spirit goes to wherever you placed your faith.

As sure as my spirit will glide away,

My soul as well, will go its own way.

Once…

Born into darkness,

But death a new day.

~ **Lloyd Joshua Sams**

Speak it to Being

August 2006

por El Ave Caminante *(by The Birdwalker)*

Chapter 9

The Apple Doesn't Fall Too Far from the Tree (The Fruit Doesn't Fall Too Far From the Root)

[22] But the fruit of the Spirit is love, joy, peace, longsuffering, kindness, goodness, faithfulness, [23] gentleness, self-control. Against such there is no law. Galatians 5:22-23

Follow, "The apple doesn't fall too far from the tree." Further, "The fruit doesn't fall too far from the roots." As far back in time as man's history goes in the pages of ancient texts, we have recorded the struggles of man attempting to contextualize God - the Creator; theorizing what He meant instead of accepting the lines in Genesis through to Revelations!

I have heard it said: That the difference between genius and insanity is the evidence of the fruit that they bear. The difference between Genesis and Revelations is the circumpunct of the cross which illuminates all our lives - The Light of the World.

If a lifetime is the equivalent of a second; a thousand years is a day in the mind of God, then that means if it took seven days to create the heavens and the earth how much more thought did he put into making man? Most people think it was in an instant – it was not. It was a process. There were steps and phases to man's coming to be: He made man. He created man. He blessed man. He formed Man.

I remember reading a Gideon Bible in a hotel room once. Somewhere at the beginning of the book's thin page, I found the words written just before the book of Genesis and after the Table of Contents listing all the individual books of the Bible's New Testament.
It read:

The Bible –

1. Read it to be wise.
2. Believe it to be safe.
3. Practice it to be holy.

And just beneath these three instructions from the Gideon Brethren it said (*which was most enlightening to me, even though I've read the Bible in its entirety several times*):

"It contains light to direct you,
Food to support you,
And comfort to cheer you.
It is the traveler's map,
The pilgrim's staff the pilot's compass,
The Soldier's sword and,
The Christian's charter."

First, it was more than one day to actually create man; it first began with a thought in the mind of God. According to Moses in the Book of Genesis the heavens and the earth was created void without form (there was darkness, depth, and waters).

Day 1: Dark and Void

- Deep / Waters
- Light was good
- Day
- Night

Day 2: Air, Land, and Sea

- Firmament divides water form land
- Separates heaven from earth
- Created dry land called earth
- Gathering of the waters called the seas

Day 3: Lights, Time, Days

- Lights in the firmament
- Stars in the heavens
- Days, seasons, years, skies

Day 4: Sun & Moon

- Created Sun – the Greater Light
- Created Moon – the Lesser Light

Day 5: Living Things

- Created Fish
- Creatures in the waters
- Birds to fly

Day 6: Beasts of the Field

- Created cattle
- Created beasts
- Created creeping and crawling things
- Created man in His own image

God divided light from dark and called light Day and darkness He called Night (so from evening to morning is the first day).

Gen. 1:26

(God's pronouncement of his burgeoning forethought)
"Let Us make man in Our image, according to our likeness; let them have dominion over the fish of the sea, over the birds of the air and over the cattle, over all the earth and over every creeping thing that creeps on the earth."

Gen. 1:27

So God created man in His own image; in the image of God created him; male and female He created them.

Gen. 1:28

Then God blessed them, and God said to them, "Be fruitful and multiply; fill the earth and subdue it; have dominion over the fish of the sea, over the birds of the air, and over everything that moves on the earth."

Gen. 1:29

And God said, "See, I have given you every herb that yields seed which is on the face of all the earth, and every tree whose fruit yield seed; to you it shall be for food."

Gen 1:30

Also, to every beast of the earth, to every bird of the air, and to everything that creeps on the earth, in which there is life, I have given every green herb for food:" and it was so.

Gen. 1:31

Then God saw everything that the he made, and indeed it was very good. So the evening and the morning were the sixth day.

*__**Notice**__: it's not until Chapter 2 of Genesis does God actually form man…see 2:7 (And the Lord God formed man of the dust of the ground and breathed into his nostrils the breath of life; and man became a living being.)

God asked for permission. | God then began to create. | God finally formed man.

See then: God's process after He created the heavens and earth to be - After the greater and lesser lights; after the stars after the fish, birds, beasts, cattle, and creepy things that crawleth upon the earth and the creatures that fill the seas – even after this whole process, starting with "Let there be light," and after all this… "let" is a word that connotes permissiveness and a verbal command that speaks to a higher authority. Why does God ask permission to make man in [Their] image? Does God care more about the animals and herbs more than he does man and woman?

Gen. 2:18

Here's your answer: And the Lord God said, "It is not good that man should be alone; I will make him a helper comparable to him." … But for Adam there was not found a helper comparable to him. And the Lord god caused a deep sleep to fall on Adam, and he slept; and He took one of his ribs and closed the flesh in its place.

Then the rib which the Lord God had taken from man He made into a woman, and He brought her to the man. And Adam said: "This is now bone of my bones and flesh of my flesh; she shall be called Woman, Because she was taken out of Man."

Therefore, a man shall leave his father and mother and be joined to his wife, and they shall become one flesh. And they were both naked, the man and his wife, and were not ashamed. As you can see, they were "naked" and not "ashamed" of their post-op cuts, marks, bruises, flaws, or natural blemishes (or even the difference in their sexual anatomy).

As you can see, it is not until the second chapter of Genesis that Adam and Eve are introduced in biblical verse. This is important because *Sunshine Came* is about a process that every individual will face in order to liken themselves unto their Creator; enlightened unto a higher state of being that is always rewarding and fruitful.

The Sun & the Moon & The Stars

Secretly stars stare silently at night,
Day drops darkness, diminishing their sight.
The sky scatters across calendars like a bi-racial chessboard.
Squares with their checkered hue like white chalk on a blackboard;
The sun scans the surface of the desolate desert's sand dunes.
The moon makes lovemaking memories, to your favorite tunes.
Morning Star scales high above shining His light,
Looming lunar rays shut eyelids tight.
The Power strip surges energy from the same cord;
Showering fowlers with song and harmonic chords,
The Sun sweeps obscurity with a Phoenix-feathered broom.
The Moon illuminates lives in the nocturnal green room.
Fluorescent flowers form pads for flies to take flight,
Blossoming budding bulbs' pedals are bright.
Clouds whet their appetite when rain is poured.
Spilt from mighty tiers of the Lord.
The Sun creates star births inside the earth's womb.
The Moon reflects mirth beside its tomb.

~ By Lloyd Joshua Sams

Vida Inteligente (Intelligent Life), 2009

por El Ave Caminante *(by The Birdwalker)*

Chapter 10
The Science of Light Bodies
(Patterns and Cycles)

The Light in each of was given by our Creator to navigate through this dark world. The Kingdom on earth is established in the realm of light; "trouble lasts but a night but joy cometh in the morning" (Psalms 30:5) We associate night with the absence of a fully lit sun, but as we learned in Genesis, the Creator made a greater light and a lesser light. The greater light, or the sun, shall rule what is called day; and the lesser light or the moon (reflection from the sun) shall rule the night.

Most humans believe that the sun moves its location at night, but scientifically, we know that the sun is Always present and never moves. The earth's orbit rotates on its axis constantly changing its directional relationship to the sun which presents light and dark during the day and night. The earth's movement around the sun changes weather patterns, seasons, and our psychological conditions.

Periodic menstruation: lunacy, and tantrums are all connected to the lunar cycle and earth's rotation, as well as flares released from the surface of the sun. The moon is a mere reflection of light from the Light of the World.

Sunlight is essential to human life and well-being. Our bodies are naturally wired to the sun and its cycles, which dictate our circadian rhythm and regulate our sleep patterns. The sun provides us with warmth, light, and energy, and has been credited with numerous health benefits, including improved mood, enhanced immunity, and increased mental alertness.

Unfortunately, many of us now spend most of our days indoors, away from the sun's nourishing rays. As a result, many people suffer from a deficiency in Vitamin D, which can lead to a range of health problems, including osteoporosis, weak immune systems, and a greater risk of developing certain types of cancer.

A human beings' metabolism and circadian rhythm regulate a person's vitality, emotional state, and physiological status. Stress affects health status, mental health status, and activates adrenaline (adrenal gland). The cortisol gene produces natural stress relievers in the body to counteract stress. Each person has a unique metabolic signature that controls heart rate; weight gain / loss; mood; and dietary appetite as well as sleep.

Isaac Newton is one of the most renowned scientists in history, known for his groundbreaking discoveries in mathematics, physics, and astronomy. One of his most significant contributions was his study of light and color. Newton's work on light laid the foundation for our modern understanding of the properties of light and how it interacts with matter.

In 1666, Newton conducted a series of experiments with light that led him to conclude that white light is composed of all the colors of the rainbow. He demonstrated this by passing a beam of sunlight through a prism, which separated the light into its component colors. Newton called this a spectrum, which is the range of colors that make up white light. He then conducted further experiments to study the properties of light and color.

Newton's experiments led to the development of the theory of color, which explains how light interacts with matter to produce color. He discovered that the colors in the spectrum were always the same, and that they were determined by the properties of the light itself. For example, red light has a longer wavelength than blue light, which is why it appears at one end of the spectrum and blue at the other end.

Newton also discovered that light can be refracted, or bent, when it passes through a transparent medium, such as a prism or a lens. This led to the development of lenses for eyeglasses and telescopes, which revolutionized the fields of optics and astronomy.

Moreover, Newton was able to demonstrate that the color of an object is determined by the way it absorbs and reflects light. For instance, a red apple appears red because it absorbs all the colors of the spectrum except for red, which it reflects back to our eyes.

Overall, Newton's discoveries about light and color revolutionized the field of optics and had a profound impact on our understanding of the world around us. Without his groundbreaking work, we would not have the advanced technologies we use today, such as cameras, telescopes, and lasers. His legacy continues to inspire scientists and engineers to explore the properties of light and its applications in modern society.

Every second, minute and hour of each day is connected to a specific rhythm connected to your unique individual frequency.

When you walk through an airport electromagnetic security scanner, they can see your light body; MRIs and CAT Scans can see your light body.

Three Royal One Watcher, 1999

por El Ave Caminante *(by The Birdwalker)*

Chapter 11

We Are Seeds Planted in God's Garden

The growth of a seed and plant can be compared to the process of human procreation and gestation. Both involve cycles, the need for sunlight and water, and the importance of male and feminine energies. Water is the catalyst for life.

In the case of plants, the process begins with the seed, which contains all the genetic information necessary for growth and development. Just like human procreation, the male and feminine energies come together to fertilize the seed, initiating the growth process. From here, the plant requires sunlight and water to grow, just as the human fetus requires nourishment and care from the mother.

Different types of plants require varying amounts of sunlight and water to grow. For example, orchids require a specific balance of light and shade, as well as moisture and humidity, to thrive. Similarly, lotus flowers grow best in shallow, warm water and require ample sunlight to bloom. Venus flytraps, on the other hand, require a moist soil and plenty of sunlight to catch their prey.

Seasonal plants also have unique requirements for growth. For example, spring flowers like tulips and daffodils require plenty of sunlight and moderate watering to thrive, while winter plants like holly and poinsettias prefer cooler temperatures and less water. Nocturnal plants, such as night-blooming jasmine, have evolved to attract pollinators like moths and bats that are active at night.

Just as the growth of a plant requires water, sunlight, and male and feminine energies, so does the process of human procreation and gestation. The female reproductive system provides the nourishing environment necessary for the fetus to grow, while the male contribution of sperm provides the necessary genetic material for development. Similarly, the process of conception and gestation requires the balance of male and feminine energies, just like the balance of light and water for plant growth.

It is also interesting to note the role of fragrances in both plants and humans. Lavender, jasmine, and bergamot are all plants that have been used for their fragrance for centuries. These fragrances have a calming and soothing effect on humans and are often used in aromatherapy. Similarly, humans also produce pheromones, which are chemical signals that play a role in attracting potential mates.

The growth of a seed and plant can be compared to the process of human procreation and gestation in several ways. Both involve cycles, the need for sunlight and water, and the importance of male and feminine energies. Different types of plants have unique requirements for growth, just as humans have different needs during pregnancy. The use of fragrances in both plants and humans further emphasizes the similarities between the two.

The Role of Sunlight in Plant and Human Pigmentation

Sunlight provides a multi-color spectrum of light that is essential for both plants and humans to develop pigmentation. Pigments are molecules that give color to living organisms, and without sunlight, they would not be able to develop properly. In plants, pigments such as chlorophyll and carotenoids are responsible for the green and yellow colors we see in leaves and fruits. In humans, melanin is the pigment responsible for skin, hair, and eye color.

The production of pigments in plants and humans is a complex process that requires the absorption of specific wavelengths of light. Chlorophyll, for example, absorbs mostly blue and red light, while carotenoids absorb mostly blue and green light. Similarly, melanin production in humans is regulated by exposure to ultraviolet (UV) light from the sun. However, exposure to too much UV light can also be harmful, causing skin damage and increasing the risk of skin cancer. That is why it is important to find a balance between getting enough sunlight to produce pigments and protecting oneself from too much UV exposure.

Photosynthesis

Photosynthesis is a biological process that occurs in plants and some microorganisms. It is the process by which plants convert light energy from the sun into chemical energy that they can use to fuel their activities. This process is essential for the survival of plants and, ultimately, all living creatures on Earth.

Photosynthesis involves a complex series of chemical reactions that occur within plant cells. It all starts with the absorption of photons, or particles of light, by specialized pigments called chlorophyll that are found in the chloroplasts of plant cells. These pigments are able to absorb photons from specific wavelengths of light, which allows plants to capture energy from the sun.

Once the photons have been absorbed by chlorophyll, they are used to power a series of chemical reactions that convert carbon dioxide and water into glucose and oxygen. The carbon dioxide is obtained from the air through tiny pores on the surface of the leaves called stomata. The water is absorbed by the roots of the plant and transported to the leaves through specialized tubes called xylem.

The chemical reactions involved in photosynthesis are very complex and require a lot of energy to occur. This energy is provided by the photons that are absorbed by the chlorophyll. The process of photosynthesis is divided into two stages: the light-dependent reactions and the light-independent reactions.

During the light-dependent reactions, the photons that are absorbed by the chlorophyll are used to produce a molecule called ATP, which is used to power the light-independent reactions. In addition, the light-dependent reactions produce a molecule called NADPH, which is used as a reducing agent in the light-independent reactions.

The light-independent reactions, also known as the Calvin cycle, use the energy stored in ATP and NADPH to convert carbon dioxide into glucose. This process is very important because glucose is the primary source of energy for all living organisms. The oxygen that is produced during photosynthesis is released into the atmosphere as a waste product.

The transmission of photons through wavelengths is similar to the way that television signals or the internet are transmitted through fiber optic cables. In both cases, information is transmitted using electromagnetic radiation. In the case of photosynthesis, the photons are transmitted through the electromagnetic spectrum and absorbed by chlorophyll in plant cells.

The electromagnetic spectrum is the range of all possible frequencies of electromagnetic radiation. This includes everything from radio waves and microwaves to X-rays and gamma rays. The visible spectrum, which is the part of the spectrum that we can see, ranges from approximately 400 to 700 nanometers.

Different pigments in plants are able to absorb photons from different parts of the electromagnetic spectrum. This is why plants appear green to the human eye, because chlorophyll absorbs photons in the red and blue parts of the spectrum but reflects photons in the green part of the spectrum.

In summary, photosynthesis is a complex process that allows plants to convert light energy from the sun into chemical energy that they can use to fuel their activities. This process is essential for the survival of plants and, ultimately, all living creatures on Earth. The transmission of photons through the electromagnetic spectrum is crucial to this process, allowing plant cells to absorb the energy they need to power the chemical reactions involved in photosynthesis.

The Aesthetic Beauty of Sunlight in Plants and Humans

Sunlight also plays a significant role in the aesthetic beauty of both plants and humans. In plants, the angle and intensity of sunlight can influence the shape and texture of leaves and flowers. For example, a plant that receives more sunlight on one side than the other may grow leaves that are angled towards the light source, resulting in a unique shape.

Similarly, in humans, sunlight can also influence physical appearance. Exposure to sunlight can result in a healthy and glowing complexion, while a lack of sunlight can result in a pale and dull appearance. In addition, sunlight can also affect hair color, with exposure to UV light resulting in a lighter shade.

Sunshine also plays a significant role in the beauty industry. Tanning beds and UV lamps are commonly used to achieve a desired skin tone, while light therapy is used to treat a variety of skin conditions, such as acne and psoriasis. In the plant world, horticulturalists and landscapers use sunlight strategically to create beautiful gardens and landscapes.
In conclusion, sunlight plays a vital role in both the functional and aesthetic aspects of plant and human life. The multi-color spectrum of sunlight is essential for the production of pigments that give color to plants and humans, while also influencing their physical appearance. The proper balance of sunlight exposure is crucial for optimal health, beauty, and growth.

The concept of showing love and affection towards plants and water is not new. In ancient times, various cultures considered plants and water as living beings, and they showed reverence and respect towards them. With the advent of science, research studies have shown that plants and water indeed have life and can respond to human emotions and words. This research report aims to explore the scientific studies that have investigated the effects of positive messaging, love, and affection on plants and water, their responses to external stimuli, and their growth.

Water and Memory:

The concept of water having memory is a relatively new phenomenon, and several studies have shown that water can indeed store and transmit information. One study conducted by Masaru Emoto, a Japanese scientist, examined the molecular structure of water when exposed to positive or negative emotions, music, and words. The study found that water exposed to positive words and emotions had a more symmetrical and beautiful molecular structure than water exposed to negative emotions and words.

The study also revealed that water could retain this structural change for an extended period. Another study conducted by Bernd Kroplin, a German scientist, examined the memory of water and its ability to store and transmit information. The study found that water molecules could store and transfer information much like a computer's hard drive.

Water has been found to have the ability to store information and memories, a property that has been the subject of scientific studies for several years. In the 1980s, Dr. Masaru Emoto, a Japanese scientist, conducted experiments that suggested water can carry information in the form of vibrations. Dr. Emoto exposed water to different words, music, and images and then froze the water, which allowed him to observe its crystalline structure under a microscope.

Dr. Emoto found that water exposed to positive words and emotions formed beautiful, intricate crystal patterns, while water exposed to negative words and emotions formed distorted, incomplete crystals.

In another study, researchers found that water exposed to human DNA produced a different crystalline pattern than water exposed to other sources of DNA. This suggests that water can "read" and store genetic information. The implications of this research are significant, as it suggests that water may play a role in the transfer of genetic information between cells.

I postulate that every molecule has consciousness. And that this postulation will be proven post publishing this book.

Plants' Sensitivity to External Stimuli:

Plants are living beings that can sense and respond to the world around them. Several studies have shown that plants can detect changes in light, temperature, sound, and touch. One study conducted by Monica Gagliano, an Australian scientist, found that plants could sense and respond to gravity.

In the study, Gagliano observed that plants grown in a slanting position would grow in the direction of gravity, even if it was at an angle. The study concluded that plants could detect the direction of gravity and adjust their growth accordingly.

Positive Messaging to Plants:

Several studies have investigated the effects of positive messaging to plants, and the results have been astounding. One study conducted by Dr. Masaru Emoto examined the effect of positive and negative words on the growth of rice plants. The study found that plants exposed to positive words grew significantly taller and healthier than plants exposed to negative words.

Another study conducted by Dr. Theodore Barber examined the effects of positive messaging on the growth of beans. The study found that plants exposed to positive messaging grew significantly taller and had more extensive root systems than plants exposed to negative messaging.

Cellular Memory and Unlimited Growth

As a child who grew up with scales like a fish because of my psoriatic skin cancer, I often wondered why I shed so much DNA from my skin. Anyone with psoriasis knows that the replication of your skin cells results in shedding millions of skin cells a day.

As a therapy, I was a PUVA patient, which meant I was placed in a light chamber that emit artificial UV rays at radionic levels to remedy and retard the growth of skin cells so that my body would stop shedding skin. The therapy never cured my condition, however, it helped tremendously. Like the story of Hansel and Gretel reminded me that perhaps God intended for me to shed my skin so that I may be able to map my travels around the sun.

More recently, after studying as a Research Fellow at The University of Texas M.D. Anderson Cancer Center, I learned that cells in our bodies have the ability to store information and memories as well. Research has shown that cells have a "memory" of their past experiences, which can influence their behavior and function. For example, cells that have been exposed to a particular chemical or stimulus may become more responsive to that stimulus in the future.

One example of cellular memory is the case of Lackey cells, which are cells that have been cultured from a cancer patient. These cells have the ability to grow and divide indefinitely, a property that is not found in normal cells. Researchers have found that Lackey cells have a unique genetic code that allows them to bypass the usual limitations on cell growth and division. This genetic code may be a form of cellular memory that allows the cells to remember their ability to grow and divide without limit.

In addition to genetic memory, cells also have the ability to store information in their skin or outer layers. This can include information about the environment, such as temperature, pressure, and chemical signals. Cells can also store information about the past experiences of the organism, such as exposure to toxins or disease. This cellular memory can influence the behavior and function of cells and can have important implications for health and disease.

The studies discussed in this research report provide evidence that plants and water are living beings that can sense and respond to external stimuli, including human emotions and words. The studies have shown that water has memory and can store and transmit information, while plants can detect changes in light, temperature, sound, and touch.

Moreover, the studies have shown that positive messaging, love, and affection towards plants and water can have a significant impact on their growth and well-being. These findings suggest that showing love and affection towards plants and water can increase harmony and healthy growth, and provide a more profound connection between humans and nature.

Citations

1. Emoto, M. (2004). The Hidden Messages in Water. Atria Books.
2. Rey, L. (2003). Thermoluminescence of ultra-high dilutions of lithium chloride and sodium chloride. Physica A: Statistical Mechanics and its Applications, 323, 67-74.
3. Upadhyay, R. P., Nayak, C., & Singh, R. H. (2011). Effect of homeopathic remedies on breast cancer cells. International Journal of Oncology, 36, 395-403.
4. Kiecolt-Glaser, J. K., Glaser, R., Gravenstein, S., Malarkey, W. B., & Sheridan, J. (1996). Chronic stress alters the immune response to influenza virus vaccine in older adults. Proceedings of the National Academy of Sciences, 93, 3043-3047.
5. Lin, S. J., Defossez, P. A., & Guarente, L. (2000). Requirement of NAD and SIR2 for life-span extension by calorie restriction in Saccharomyces cerevisiae. Science, 289, 2126-2128.
6. Shay, J. W. (2001). Role of telomeres and telomerase in aging and cancer. Cancer Research, 61, 3509-3511.

Chapter 12

Frequencies & Waves

Waves are a fundamental concept in physics and refer to the oscillations that propagate through space and time, carrying energy and information. Waves are ubiquitous and manifest in various forms, from electromagnetic waves to oceanic waves, sound waves to seismic waves. In this essay, we will explore the different types of waves and frequencies, their characteristics, and their significance in everyday life.

One of the most common types of waves is the electromagnetic wave, which is a form of energy that travels through space as electric and magnetic fields oscillating at right angles to each other. Electromagnetic waves are categorized according to their frequency or wavelength, which determines their properties and applications. For example, radio waves have the longest wavelength and the lowest frequency and are used for communication and broadcasting, while X-rays have the shortest wavelength and the highest frequency and are used in medical imaging and radiation therapy.

Another type of wave is the oceanic wave, which is a disturbance that propagates through the ocean's surface due to wind, gravity, and other factors. Oceanic waves are classified based on their wavelength, frequency, and height, and they can have a significant impact on coastal regions, marine ecosystems, and human activities such as shipping, fishing, and tourism.

Sound waves are another example of waves that are essential for communication and perception. Sound waves are longitudinal waves that travel through a medium such as air, water, or solids, and are characterized by their frequency, wavelength, amplitude, and velocity. Human hearing is limited to a range of frequencies between 20 Hz and 20,000 Hz, but animals such as bats and whales can detect higher frequencies and use them for navigation and communication.

In addition to physical waves, there are also social and cultural waves that influence human behavior and trends. For example, fashion trends, musical genres, and political movements can be seen as waves that emerge, peak, and decline over time, driven by factors such as innovation, popularity, and social norms.

The cyclical nature of waves is also evident in natural phenomena such as weather patterns, seasons, and geological processes. For instance, the rotation of the earth around the sun creates a cycle of day and night, while the tilt of the earth's axis causes the four seasons to occur. Geological waves, such as seismic waves, can reveal the structure and composition of the earth's interior and provide valuable insights into geology and natural hazards.

Frequencies and wavebands are essential concepts in wave theory, and they refer to the number of oscillations or cycles that occur in a second. The unit of frequency is the Hertz (Hz), which corresponds to one cycle per second. Different frequency ranges are associated with different applications and phenomena. For example, the human brain emits electromagnetic waves in the range of alpha waves (8-13 Hz) and beta waves (13-30 Hz), which are associated with relaxation and concentration, respectively.

In the field of electronics and telecommunications, high and low-frequency wavebands are used for different purposes. For example, high-frequency bands such as microwaves and infrared waves are used for remote sensing, communication, and navigation, while low-frequency bands such as AM radio and shortwave radio are used for broadcasting over long distances.

Waves and frequencies are pervasive in nature and society, and their study has led to significant advances in fields such as physics, engineering, and medicine. Waves and frequencies are essential for communication, perception, and navigation, and they reveal the underlying patterns and structures of the world around us. By understanding the properties and characteristics of waves, we can better appreciate their significance and harness their potential for innovation and progress.

An antenna is an electrical device that converts radio waves into electrical signals, and vice versa. It works by receiving electromagnetic waves from a transmitter and converting them into an electrical signal that can be processed by a radio or television receiver. The size and shape of the antenna determines the frequency of the waves it can receive.

Radio waves are a type of electromagnetic radiation that are used for broadcasting radio and television signals. These waves have a long wavelength and low frequency, which allows them to travel long distances without being absorbed by the atmosphere. They are divided into two types of frequencies: AM (Amplitude Modulation) and FM (Frequency Modulation).

AM radio waves have a lower frequency than FM radio waves, which makes them more resistant to interference from buildings, hills, and other obstacles. They can travel long distances by bouncing off the ionosphere, a layer of charged particles in the Earth's upper atmosphere. However, their lower frequency also means that they have a lower sound quality than FM radio waves.

FM radio waves have a higher frequency than AM radio waves, which allows them to transmit higher quality sound. However, they have a shorter range and are more easily absorbed by obstacles such as buildings and hills. FM radio waves are typically transmitted by line-of-sight, which means that the signal is strongest when there are no obstacles between the transmitter and receiver.

The quality of radio and television signals can be affected by environmental factors such as sunlight and clouds. Sunlight can cause interference with radio signals, especially during the day when the ionosphere is charged by the sun.

This can lead to a decrease in the quality of radio broadcasts, and sometimes even cause them to be completely disrupted. Clouds can also affect radio signals, as they can absorb or reflect the waves and cause interference.

To improve the quality of radio and television signals, broadcasters use a variety of techniques to reduce interference and improve transmission. For example, they may use directional antennas that focus the signal in a specific direction, or they may use repeaters or boosters to amplify the signal in areas where it is weak. Additionally, broadcasters may use satellite or cable transmission to bypass atmospheric interference altogether.

Chapter 13

Travel Light

When we hear the phrase "traveling light," we often think of packing only the essentials for a trip and leaving behind the unnecessary items. However, the concept of traveling light, here, can also apply to the weight we carry in our hearts and minds as we journey through life. Just as heavy luggage can make a trip more cumbersome; the weight of emotional baggage can weigh us down and prevent us from experiencing the joy and beauty of life.

One way to travel light is to let go of grudges and resentments. Holding onto anger and bitterness takes a toll on our emotional well-being and can hinder our relationships with others. Forgiving those who have wronged us frees us from the burden of carrying around negative emotions and allows us to move forward with a lighter heart.

Another way to travel light is to focus on the present moment. Often, we get caught up in worrying about the future or dwelling on past mistakes. This mental baggage can prevent us from fully experiencing the beauty of the present moment. By staying mindful and focused on the present, we can let go of unnecessary worry and enjoy the journey.

Traveling during the day is also safer than traveling at night. When we can see clearly, we are less likely to stumble or get lost. Similarly, when we approach life with a clear mind and heart, we are better equipped to navigate the challenges and obstacles that come our way.

Taking the metaphor of traveling light a step further, we can also think of the Son of God watching over us as we journey through life. Just as we might feel more at ease traveling with a trusted friend or companion, knowing that God is with us can provide a sense of comfort and security. When we trust in God's plan for us, we can let go of fear and worry and embrace the journey with a sense of peace and hope.

In essence, traveling light means letting go of the things that hold us back and weigh us down. By focusing on the present moment, practicing forgiveness, and trusting in a higher power, we can journey through life with a lighter heart and a greater sense of purpose.

We are all candelas moving about our planet, waxing and waning, dancing as fireflies during a summer night. In our troupe, we form alliances and congregate like schools of electric eels marauding wisdom through deception using dark tactics, while the path to the kingdom is within.

No matter how far we rove in this world we are never distant from God; we're connected to the Alpha and have always been tied to his infinite plan prior to our awareness of consciousness or spiritual experiences.

In the absence of Sunshine, we should remember that: The Sun isn't missing or gone - we are simply transitioning to a new place in our process to be more mature beings - enlightened to a higher consciousness; a godly or angelic or saintly mindset, rather than our 21st Century evolved contemporary thought process about, heaven, the afterlife and eternity.

When it rains or pours; when TDC's (thundering dark clouds) storm into your life; when shadows seem to outnumber clear images; when your clarity and judgment seem cloudy and confusing know that the closer you seek or draw to the Light of the World, the closer the Light of the World will draw to you (James 4:8).

Tragedy can strike at any time. Calamitous situations can be presented at a moment's notice, How will you react it hits you? Are you going to fold? Can you stand up to the pressure? Or will you remember that trouble doesn't last always: "For his anger lasts only a moment, but his favor lasts a lifetime; weeping may stay for the night, but rejoicing comes in the morning. (Proverbs 30:5 NIV)

When in the flesh, Jesus Christ was crucified (terrible); but He knew God's plan - the Main Objective, His purpose for being on earth - His higher calling beyond the natural and into the supernatural which is everlasting. Unlike Jesus Christ, we meander most of our lives without living without fulfilling our purpose. Jesus showed us that there is more than meets the eye here on earth.

In case you're wondering what is your purpose? - it is to follow the Light of the World; to Love; to cast your light as a lampstand, not to be just a wisp of smoke blowing in the wind, going hither and thither.

Sunshine Was His Name

Every day he comes to visit
And peeps through my window;
Sneaks between the blinds
He passes his orange lips upon my skin.
I feel his warmth passing in,
Through my body,
Like water swishing
Thru a flounder's gill,
Or like wind whipping
And flapping a half-staved flag.
My eyes peel, away from darkness,
And they too, acquaint, themselves
With his mysterious glowing appeal -
Oh, I would love to feel him; stroke and touch him.
But I cannot.
I fiend for his voice; it's pleasant as a siren by the sea,
But as he lay next to me,
My retinas gaze-
Like x-rays, thru his translucent physique.
All I can sense is his warmth.
Like the Earth saturates the morning dew,
So does the air swallow his voice,
I motioned to inhale his lovliness,
And by just wishing for a kiss-
Suddenly! He went away.
He hid behind a tree, and then a cloud –

Or maybe 2 or 3 –

At dawn, he retreats from his Nemesis – the Moon…

Zoom, zoom, like a sonic boom!

He disappeared and reared behind the Horizon,

Passing infinite space;

He's gone now and I miss his shining face.

"Wait!" I yelled up to the sky…

"When can I see you again?"

But no reply –

No answer came back,

Until dusk cracked Day

Over the Eastern Bay -

Just above where my head lay.

Sunshine is His name,

And forever shall he remain!

~ Lloyd Joshua Sams (09 / 09 / 1998)

Chapter 14

Final Thoughts

(The Sunshine is Coming Again)

Letting your light shine involves the highest responsibility, "it will reward the greatest labor, and will condemn all who trifle with its sacred content." (Gideon's Prayer) As a lampstand, you are a beacon of light to those in darkness. As your spiritual IQ grows, your power to wield the lancing sword of Light (Truth, Love, and Wisdom) will become like muscle memory or the "phantom memory;" holding the key to enlighten others will become natural rather than supernatural.

On earth we are held captive and can choose Faith to free us from this bondage of trickery, allusion, and material greed. Heaven is a place filled with "light beings," or Children of Light seeking to cast down rays of goodness and insights to us as we sojourn toward our purpose on this planet, or wherever we find ourselves under the auspice of The Ancient of Days – Alpha & Omega.

It has been said for centuries that the eyes are the windows to the soul. The eye collects light rays and processes signals and frequencies of refracted and reflected wavelengths into images; making logical sense of the physical world around us: proximity, color, danger, food, and safety. Remember, seek first the kingdom of Heaven <u>within</u> you; eyes are the peep holes to Heaven – the source.

We may not (currently) be able to transfigure ourselves into "a light body" like Jesus did at Mount Sinai, but we can share our light by allowing others to see the Light of the World through our countenance (both the greater and lesser lights as described in Genesis). A warm inviting smile: a genuine "Hello," a simple complimentary glance can make a person's day, whereas a non-response; frown, or avoidance accomplishes nothing of importance.

"Do not judge, or you <u>too</u> will be judged. For in the same way you judge others, you will be judged, and with the measure you use, it will be measured to you. "Why do you look at the speck of sawdust in your brother's eye and pay no attention to the plank in your own eye? How can you say to your brother, 'Let me take the speck out of your eye,' when all the time there is a plank in your own eye?

You hypocrite, first take the plank out of your own eye, and then you will see clearly to remove the speck from your brother's eye."
Matthew Chapter 7:1-5

Humans perceive reality based on 90% of what we <u>see and hear</u> and 10% from the other senses. But the problem is you can't trust everything you see: 10% of what you see is real or good for you, 90% is artificial and a distraction from obtaining the Truth. Although the brain processes "hearing" information thousands of times faster than images: the brain records more memories of images than of sound (for those who have the gift of vision). Listening is a survival skill to escape predators.

Even with eyesight, many people are still "blind" because they are lost in the dark; a pseudo prison of the mind that traps them inside of ignorant and non-progressive thinking. Our thoughts control our actions. Ultimately, **what we think can create the environment around us**.

For years the star has been a symbol of honor and majestic adornment. Like precious jewels the brilliance of a star can capture the attention of the dullest of eyes. They have been placed on Christmas trees, wrapping paper, greeting cards, in famous paintings and notably applied to coat-of-arms and government insignias. What is man's obsession with the collective glorification of the Star?

Even newborn children are fascinated with dancing lights and star trinquettes attached to their strollers and beds at night. People of all ages go to movies to see their favorite celebrity "stars" act in films portraying god-like qualities. Scientists and astronauts search the heavens for clues about human existence through studying patterns in the stars. Again, I ask, what is the impetus that drives us toward the Star(s) or the shine emitted from these dwarfing light bodies?

Are they simply something familiar from a time gone by? Do they speak to a higher connection that humans have to the Sun or their Creator? Is the star conscious? Why do we celebrate or accept "stars" over common rocks? Who placed a premium on stars over planets?

Has anyone ever put an ordinary rock or globe on top of a tree for decorations? I haven't seen it done in our culture. But I have seen Black Santa Clauses, Anglo angels, and even poinsettias - a vibrant plant or flower.

All of these images have a close correlation to the light a star gives off - The Black Santa Claus relates to minority depictions of a historically religious connection to Christ Jesus disparate from the classical or popularized notion of a European pale face lighter skinned Jesus; even though archeologists and historians know that the Sun of the world favored the peoples in the region which the lineage of Christ was said to have been born and raised between Cairo, Egypt and Bethlehem, Israel.

Semitic genes are found in bloodlines, generation after generation - phenotype and appearance resemble the description of cherubim angels seeking their Light Source thousands of years now appearing on earth as Dark or Black-faced (melanin in their skin); King Solomon said, "I am not dark because I have sinned. I am dark because the sun has favored me greatly."

One then, could safely assume that King Solomon also considered his lineage and entirely all the kingdom's subjects to also be favored by God. Much like the *"Sunshine My Sunshine"* song this idea gives credence to the point that there is a greater more powerful force, the "invisible hand" that guides things in motion - the Light of the World - Christ Jesus.

Remember, I never intended for this book to be a self-help or religious read. I bring up historical points only to highlight that *Sunshine Came* derived from my study of the Light of the World and the Sun to guide my thinking to balanced living here on earth and beyond; not necessarily "perfect living." I am content with being perfectly imperfect, made in the image of my Creator.

The journey of life is eternal no matter what religion you subscribe to. There is a universal law called: Reciprocity. In Latin, the inventors of legalese named this law: quid pro quo - "Something for Something" an exchange or return of equal value or cost. Like reciprocity, quid pro quo is relative to karma, which is the principle of causality (and the endless universal relationship of cause and effect).

The Sunshine Came is the unabridged truth of the parallels between our connection to the sun and all the diverse cultures of the world. Each of us are a bulb sewn in the ever-expanding universal garden - to shine like bright stars for the next generation to see.

The lotus flower is a popular symbol to karmic believers and other faiths because it is one of the few flowers that carries tiny seeds inside itself as it blooms every stage of its development cycle - a constant reminder that we are all a work in progress and to be encouraged that despite growing in muddy situations that life is a constant development process. The concept of karmic energy is described in the ancient Indian texts of Brihadaranyaka Upanishad. For example, at 4.5.5-6:

Now as a man is like this or like that,
according as he acts and according as he behaves, so will he be;
a man of good acts will become good, a man of bad acts, bad;
he becomes pure by pure deeds, bad by bad deeds;
And here they say that a person consists of desires,
and as is his desire, so is his will;
and as is his will, so is his deed;
and whatever deed he does, that he will reap.
– **Brihadaranyaka Upanishad**, *7th Century BC*

Archeologists have discovered more pyramids in Egypt below ground than above ground. What does this say about ancient civilizations? Well, in part it says that the apex of life and reaching the plateaus of travels take an inward and outward perspective. Excavating pathways to illumination in the soul activates enlightenment in the mind and upper chakra, opening portals to manifest in the material world.

The principles drafted herein are hinged on my personal experiences, Holy Scriptures, and wisdom imparted by other believers.

One must remember that darkness is an associate of light that cannot be discounted nor should be considered as an equalizer. The dark surrounds light, therefore, when light is expanded so too is the band of darkness around it. This is an important caveat because God created light out of darkness. I will address this in future writings.

The sun, the earth, and the human body are all interconnected in complex and fascinating ways. Through practices like meditation, visualization, and spending time in the sun and nature, individuals can tap into this connection and harness the energy and life-giving properties of these celestial bodies to improve their health and well-being. Whether you are seeking physical health, mental clarity, or spiritual connection, the sun and the earth offer an endless source of nourishment and renewal.

Thank you for joining me and many others around the world in reading this text. By doing so, you have enlightened your corner of the universe and perhaps will begin to co-create with the Light of the World - The Sunshine, the one who physically (not just as a deity), came to earth as a man and revealed himself to be a true God - a brother and a friend. Following The Sunshine daily, despite stormy conditions, expands your horizons and increases your borders until you are at peace in green pastures and still waters.

The Sunshine Came. And He is coming again! In the twinkling of an eye - no man knows the day, time, or hour. Herein, details the discovery and secrets of light living and rudimentary instructions on taming the fire in your belly; reflecting the Sun from within your soul; the power of peace and understanding that the Sun never sets but we often settle.

Don't accept mediocrity or artificiality.

The world is not flat - there are levels, dimensions, and degrees - mountain highs and valley lows. Everyone is not out to get you - there are "light bodies" watching constantly, eagerly awaiting the chance to shine on your parade rather than rain on it. Rising above adversity requires a deep intro-engagement with the higher You - that part that is connected to the Light of the World.

The Sun is not only a physical object in the sky, but also a symbol of God's power, grace, and presence. Whether it is standing still in the sky or shining with healing power, the sun is a reminder of God's miraculous and transformative work in the world.

In the absence of darkness there is light.

The Sun Reaching to Stars

The best side is beside the Western Sky,

Flirting with the horizon and kissing your eye.

Convening between the Yang and the Ying,

Pressing darkness and washing the unclean.

Parting from the East, rising morning yeast;

It keeps your time piece and approves your release.

Sunrise begets sun-sweat; Nocturnals fall to sleep.

Sunset ends their retreat; Warm-bloods work in the heat.

Every second counts days, which count; continuously

Accounting for hours that spent years…

The Rapture will fall upon all of us at the twinkling of an eye;

None of us can spy on God;

Yet his Word is our Pedagogue.

May we lodge on the same square foundation and

Hope our fellow souls to their new final destination.

~ By Lloyd Joshua Sams

Author's Note

Just before the release of this publication, on August 25, 2014, I, Lloyd Joshua Sams, was involved in a terrible accident which led to my SUV tumbling over several times on its side landing upside down on a hill with gas spewing everywhere. (I had just checked my tires and filled my tank 10 minutes prior to the rollover); while traveling from Dallas to Houston on Interstate Highway 45, a heavy traffic corridor in Texas.

I had just met with my ex-wife and my son, Joshua. I visited Dallas to give my son clothes and school supplies for his first day of kindergarten. Hours after playing games at Chuck-E-Cheese Pizza in Addison, TX, I was thrust into practicing what I preached in this book: Search for Light of the World; believe that all things work together, just darkness and light; and lastly God is in total control.

Testimony: Between mile markers 168 and 169 in Centerville, TX (midpoint between DAL and HOU), 48 hours before my accident, there were two fatal accidents killing both families inside their vehicles ejecting two children onto the road.

Tragically heartbreaking, I know.

Oddly enough, (I do not believe in coincidences) the prior passengers in both accidents, one day apart, 2 days before mine, were driving Ford Explorers - just as I had. The difference is that they perished, and my life was spared with minor injuries: swelling of the kidney, bruised ribs, contused hip, scarring and lacerations. But...I walked away from the accident. My car was completely totaled. 21 gallons of gasoline dowsed my head as I hung upside down from my seatbelt.

Seconds before the accident, I remember talking to God, and thanking him for giving me a Son; allowing me to travel with Him; giving me the talent and ability to share with the world - The Light of the World - Jesus Christ - the Sun/Son of God. I recall extending my right hand of favor to fellow travelers who were stranded on the side of the road changing tires or ran out of gas or maybe their car overheated - perhaps they were weary and pulled over to rest - no matter the situation if I saw someone in need on the road, I would pray for them.

At the very moment that I extended my hand to a stalled car traveling Northbound on the opposite side of the median - AT THAT VERY MOMENT - three of my tires exploded and I lost control but SHOUTED for the Holy Spirit to wrap me in His might and protect me that I would not perish - but if it were in his will to take me. I was ready and at peace because I had done all I could do, with the last thing leading up to that moment - giving provisions to my Son/Sun so that he may have, rather than have not.

There is much more I can say about the abundant power of The Light of the World - I am available for speaking engagements, lecture series, professorship, and motivational speaking. I also will continue writing and publishing works that support the Light Doctrine. I believe I was given a second, third, fourth - infinitely many chances because God truly wants this book to be released to the world.

Interestingly, my college class ring, cell phone, laptop, and digital camera along with my dress shirts for work. Everything inside the car was thrown out the broken windows over a crash site area of about a quarter mile. Even my shoes were thrown off my feet.

My camera, phone, and laptop which all carried some piece or another of the work *The Sunshine Came*; I repeated over and over to the paramedics and highway patrol, "Please I have to get my computer, I have to get my phone" (which had voice recordings of my thoughts when writing, my dictatorial expressions which I wanted to include into audio version of the book).

All was recovered. Even my gold ring was found in the bushes on the hill several feet away. Please understand that I am not materialistic. I am extremely thankful to be alive, grateful to the men and women who helped save my life and to finish this literary work for you to examine for yourselves.

The truth lies between the lines.

Sincerely,

Lloyd Joshua Sams

August 29, 2014

Author's Bio

Lloyd Joshua "The Birdwalker" Sams is a native from Houston, Texas. He graduated with honors from North Shore High School. He began college at the age of 14 at San Jacinto College and later began studies in Theater and Music Composition at The University of Texas at Austin where he learned stage production and piano performance. He has traveled to four continents & 8+ countries.

Sams, a consummate lifelong student continued his education at The University of Texas at Dallas gaining a BA, there he served as the Chess Club Vice President, Jazz Band Pianist and Student Government Senator. Sams earned his associate degree in advanced Spanish from La Universidad de Guanajuato in Mexico. He became a Research Fellow at The University of Texas at M.D. Anderson Cancer Center in Houston. While being a student at UT-Dallas, Sams was also requested several times by the Osteen Family to perform at special events at Lakewood Church in Houston.

Spending many years in creative management as a Producer, Writer, and Musician, Sams has always stayed true to his roots as a student of life and formal education. As a researcher, Sams has discovered much about the human body, the world, and religion. However, as a practitioner, Sams has found methods for healing the body; unlocking powers of the earth; and sharing his faith with anyone with an open mind.

Sams has several post-graduate degrees and certificates from universities and government agencies ranging from the Center for Disease Control and Prevention; The State Bar of Texas; Duke University; Universidade Federal da Bahia; Rice University and Texas Southern University.

Sams is a polyglot who fluently speaks English, Spanish, French, and Portuguese and conversational Mandarin and Russian. He is the author of the bestselling book *The Road to Redemption – A Story About Racial Healing in America*. He is a *Daytime Emmy Award* winner and has appeared on the ABC hit primetime show *The Hustler.*

Also, a recording artist, Sams has produced several albums such as: The Soul Café, The Haus of Zuri, Straight Black & White, Live At the Piano, Keys to the Kingdom, Rio and many more.

Sams is a husband, father, brother, son, uncle, and friend to many as he resides in The Heights of Space City, Texas and enjoys traveling, live music, and coaching youth and adults.

www.ingramcontent.com/pod-product-compliance
Lightning Source LLC
Chambersburg PA
CBHW070645220526
45466CB00001B/308

* 9 7 9 8 3 9 3 2 5 5 1 9 0 *